Automation Manifesto:
No matter what industry you are in, you must automate everything!

DAN ABBATE

Copyright © 2016 Dan Abbate

All rights reserved. No part of this book may be used or reproduced in any manner whatsoever without expressed written permission of the author. Address all inquiries to: www.Robotaton.com.

Limit of Liability/Disclaimers of Warranty: While the publisher and author have used their best efforts in preparing this book, they make no representations or warranties with respect to the accuracy or completeness of the contents of this book and specifically disclaim any implied warranties of merchantability or fitness for a particular purpose. No warranty may be created or extended by sales representatives or written sales materials. The advice and strategies contained herein may not be suitable for your situation. You should consult with a professional where appropriate. Neither the publisher nor the author shall be liable for any loss of profit or any other commercial damages, including but not limited to special, incidental, consequential, or any other damages.

The websites referred to in this book are not the property of Dan Abbate and may change at any time.

First Edition
For additional copies, visit:
www.Robotaton.com

ISBN-13: 978-1535296625
ISBN-10: 1535296623

DEDICATION

Dedicated to my awesome wife, Kelly, and to my little boy, Winston, who is the best little guy in the whole world.

CONTENTS

	A Quick Introduction	vii
0	Chapter Zero	1
1	The Four Steps of Applied Automation	10
2	The Process of Automation	19
3	Excuses Companies Make to Not Automate Their Processes	31
4	The Mindset of Automation	40
5	Fatal Mistakes That Will Kill Your Automation Project Dead	49
6	Automation Today: Exploring the Cutting Edge and its Implications	56
7	The Future of Automation	72
8	No More Options	82
9	A Quick Start Guide to Automation	88
	Afterward	94
	References	98

A QUICK INTRODUCTION

Let's get directly to the point: in a business world made up of people and process, people are the weak link. This may sound like a broad generalization, but take a moment to consider your own business, its inevitable bottlenecks, redundancies, and common problems. Are these more often the product of a failure of a business process? Or rather issues with process execution by human workers?

In today's global business environment—where nearly every product and service is being reduced to commoditized levels and true competitive advantages last for a limited amount of time—there is no room for weak links if a business is to survive and prosper. Technology must be used wherever possible to remove human labor—and human error—from business processes. In fact, I contend that an automated future is inevitable if companies are to succeed. Human labor should be used sparingly and, crucially, only on those tasks in which humans are uniquely able to outperform machines. As automation increases, that will become the only cost effective, efficient, and competitive way to use human labor. Any other use of human labor will be akin to running a business today without the internet and phone: a waste of time, money, human talent, and available solutions. If such a company survives at all, it surely will not thrive.

I called this book a Manifesto because its purpose is a call to action. A call for both business leaders and the general society to take steps <u>now</u> to prepare for an automated future. The subjects I cover are not meant to be an exhaustive survey of each topic, rather an entry point that starts the reader thinking about how an automated future will impact business, society, politics, economics, and individuals throughout the world. It's very easy to find in-depth scholarly investigations of these individual subjects and I highly suggest using this book as a guide to topics that you should further explore. Think of this book as an introduction to the important and complex issues the world will be taking on in the years and decades ahead and a head start in developing your own views on the subject.

Whether this book simply inspires you to a deeper study of the issues, prompts you to become engaged with the societal or educational changes that an automated future will require, or leads you to consider automating your own business processes in order to stay relevant in the changing marketplace, this project will be a success. As you read, look for those opportunities to dig deeper and participate in the much needed global discussion of what to do before, during, and after automation of our world over the short and long term.

DAN ABBATE

Automation is coming; it is our responsibility to prepare.

DAN ABBATE

0 CHAPTER ZERO

The natural progression of human evolution will result in continued automation of our society.

Let's start at the very beginning. In this chapter, we're going to take a look at evolution—first in the sense we're all most familiar with: biological evolution. Once we have a grasp on the processes of evolution in this context, we'll move on to the development of human culture, again through the prism of evolutionary theory. What makes humans and their culture special among all other creatures? And are the evolutionary processes that shaped us back then still shaping our culture (including our economic ventures) today?

At the most basic level, sustaining life and "surviving" requires the efficient acquisition and retention of needed energy resources by individual members of a species. Contemporary humans, like all living creatures, are the product of millions of years of evolution in varied environmental conditions and the gradual process of natural selection for beneficial adaptive traits. Unlike every other animal on the planet, though, our ability to acquire energy for survival has evolved beyond what we can harness directly via our own biology. We are now, and have been for many thousands of years previously, on an invention binge that has put a layer of technology, society, and civilization between us and biological evolutionary survival. The goals, processes, and mechanisms are the same as they have been since the beginning of life on Earth. It's just that the execution of survival for us is uniquely human.

We have understood the biological processes of evolution since the time of Charles Darwin, an understanding that has grown exponentially as we've learned more about the related sciences of biology, genetics, epigenetics (the study of environmental inputs that trigger genetic processes), and, critically, advances in computer science, which allow us to build complex genetic models and process large sets of data. But our understanding of how humans develop and evolve the non-biological processes that allow our species to flourish is less well understood. In his landmark work of evolutionary theory, *The Selfish Gene*, the biologist Richard Dawkins explored the work of evolutionary processes within human culture. Are ideas and cultural developments shaped by the same processes of natural selection? Do these ideas "reproduce" based on their usefulness to survival? Dawkins (and many scientists and philosophers since) concluded that they do and provided a framework for studying their evolution (Dawkins, 1976). We will explore some of these ideas (and extensions on this premise proposed by other thinkers) as we move forward. But for the time being, I'm asking you to entertain the idea that there is an analogous process of cultural/technological development that mirrors the evolutionary model of biological development. But unlike biological evolution, which takes thousands of years of subtle genetic change and natural selection, cultural, technological, and economic evolution happens on a much faster timescale, observable to all of us if we take the time to look.

Evolution (a quick review)

"Biological evolution, simply put, is descent with modification" ("Evolution 101," n.d.). This means the first key tenet of evolution is that every current species derives from a previous one and so on down the tree of life to the very beginning. All living things are related to each other, to some degree. We all have the same origins and by going back far enough this can be proven because we all share much of the same DNA.

Species evolve through natural selection. Environmental variables and competition from other members of the same species and dissimilar species ensure that only the fittest will be able to survive to reproductive maturity, mate, and pass on copies of their adaptively superior genes. Resulting generations are thus better equipped to survive to reproductive maturity in their own time and, if they can prevail over competition, will continue to become better adapted in subsequent generations. Surviving species will continue to flourish and adapt until changes to environmental or competitive circumstances prevent successful reproduction. At this point, unadaptive species go extinct and new, more adaptively fit species take their place in the ecosystem. Only the survival of the fittest ensures that a species remains strong and viable over time (Palmer, 2009).

Sometimes, the evolutionary process will hit a surprise "fork in the road" caused by a mutation within the DNA of an individual (and its subsequent offspring) that results in a sustained competitive advantage (Palmer, 2009). This advantage naturally selects for individuals with the mutation and against those who don't, passing on the more adaptive trait and, by preventing successful reproduction, gradually eliminating the less adaptive. Eventually, these processes lead to the extinction of the originating species or, at the very least, a split into distinct species. The randomness of the mutation process is needed to introduce new potential solutions to survival into the evolutionary process. Mutation is what drives life to evolutionarily create such a diverse taxonomy of plants and animals over long periods of time.

Primordial Soup (in the beginning)

In order to grasp what it really means to be "the fittest," one must begin the conversation in the primordial soup where life began. "If we follow the straightforward reasoning that traits common to both bacteria and archaea are inherited from a common ancestor, whereas traits that differ substantially probably evolved independently at a later stage, we can draw simple conclusions about 'LUCA' – the Last Common Ancestor of cells" (Lane, Allen, and Martin, 2010). Although it will be impossible to ever pinpoint LUCA to a specific time by tracing the tree of life backwards, we know it exists in earth's history of life somewhere. As the Earth's environment cooled, conditions ordered themselves in a way that allowed specific chemical transformations to begin. These transformations were the beginning of the metabolism process here on Earth. Metabolism is necessary to life mainly because:

> …(t)he chemical reactions of metabolism are organized into metabolic pathways, in which one chemical is transformed through a series of steps into another chemical, by a sequence of enzymes. Enzymes are crucial to metabolism because they allow organisms to drive desirable reactions that require energy that will not occur by themselves, by coupling them to spontaneous reactions that release energy. (Metabolism, n.d.).

This earliest proto-metabolic process allowed the first single-celled organisms to form between three and four billion years ago.

The chemical reactions necessary for life require energy (Palmer, 2009). There is, of course, the original energy source: the sun, which blankets the earth in energy-rich radiation. The sun's energy becomes a commodity to these life-beginning chemical reactions. Early life, through its metabolism process, begins trading energy. Those reactions that are best adapted to the environment are able to gain the energy they need to continue to grow and flourish, while reactions with weak ability to gather their required energy will not continue. In short, these original chemical reactions of life were the first to begin the evolutionary process of "survival of the fittest." Couple the evolutionary process based on successful acquisition of energy with the conditions of Earth's environment at this time and you end up with only those chemical reactions best suited to trading energy under those conditions surviving. These superior reactions then mix with other successful reactions to create new, even stronger and more efficient reactions. Repeat this same process an unimaginable number of times, all across the Earth, and finally biological life eventually develops into the recognizable form from which the tree of life begins.

Human Non-Biological Evolution

As mentioned above, the biological evolutionary process cannot be observed directly by individuals because of our short time horizon. But what about evolutionary processes (mutation, adaptation, selection) that are at work in non-biological areas? In *The Selfish Gene*, Richard Dawkins proposed a theoretical framework for approaching this question, coining the term "meme" as an analogue to the gene in biological evolution. Like our genes, memes rely on their human hosts for replication and transmission. Memes are also subject to mutation, as they are changed by one host or by the process of transmission, and the adaptive advantages and disadvantages of these mutations lead to successful replication or extinction, depending on their usefulness (Dawkins, 1976).

Sharp observation can capture many examples of this evolutionary system of information/technology adaptation in our social systems, languages, customs, and economic systems. Whether or not adaptation occurs in ideas or technology in precisely the same way it does for biology is debatable. But I believe the value of analyzing these changes in evolutionary terms—where mutation, competition, and selective pressure MUST lead to evolutionary change—is extremely useful. It's a cornerstone of what I do on a daily basis as I work with companies to streamline business processes, promote efficient automation, and reduce human labor on tasks that don't really require human skills. From the goals of survival in business and in life, to adaptation and creation of new technology to gain competitive advantage, to our economic system (which is based on money, a symbolic store of value with scarce supply to facilitate the transfer of needed resources indirectly and efficiently over time). All of these systems conform to the evolutionary process and are an expression of that process at the macro-observable level.

The Human Evolutionary Difference

Human's approach to survival of the fittest is unique among all previous and current species in history:

> ...once a settled, agricultural lifestyle had been adopted – a change that took place independently in several centers worldwide in the period between about eleven and seven thousand years ago – the rules changed entirely. A species that had formally lived in tiny numbers, and had been integrated into environments it occupied, found itself for the first time in opposition to Nature. (Tattersall, 2010)

Humans banded together to ensure their survival through societal living and worked to transform their environment to make it more hospitable to them (rather than the other way around, as is the case with essentially all other life). Humans created ever advancing technology to supplement their own evolutionary process. If they needed something to dig in the dirt to survive, they did not have to wait thousands of years to evolve an appendage specialized for this environmental necessity; they simply invented and utilized a tool–i.e. the shovel. The neuroscientist V.S. Ramachandran provides another example of this difference by contrasting ancient humans with polar bears as both attempted to live in the harsh Arctic environment. Whereas the polar bear was forced to adapt a specialized fur coat over thousands of generations, human creativity allowed our ancestors to recognize that hunting a polar bear and wearing its skins as protection would dramatically shortcut the evolutionary process. And what's even better: passing on this cultural adaptation to subsequent generations requires only a lesson on how to hunt and sew clothing (Ramachandran, 2011). When you grasp this difference, it's no mystery why humans have achieved what they have.

Commerce, the process of buying/selling/trading, further accelerates this process of "faux-evolution": I don't have all the things I need to survive, but rather than die or evolve biologically to transcend the need, I can simply get it from someone else via trade. Throughout history, humans have exchanged goods and services constantly, with processes that improve exchange thriving and less effective processes mutating or dying out. In the early years, the focus was primarily on trading items for direct survival (i.e. tools, shelter, food, clothes, etc.). But soon it became necessary to value and exchange things of unequal or of unreasonable logistical complexity. Thus, as a species, we developed the intellectual concept of money as a store of value. This facilitated complex forms of trade and rapid growth of civilizations, giving our species an "unnatural competitive advantage" compared to every other species to ever walk the Earth. Now, as a species, there were mechanisms in place to guarantee the survival of individuals outside of their strict biological fitness.

Technology as an Expression of Human Societal Change

Where Dawkins and other thinkers have proposed theories about the processes of technological and cultural evolution (i.e. the evolutionary model of change and natural selection for utility), Richard Florida, Director of the Martin Prosperity Institute at the University of Toronto and Global Research Professor at New York University, asserts that technological developments are often direct expression of human societal changes. In a recent article for the *Chronicle of Higher Education*, Florida writes:

> The dustbin of history is littered with dire predictions about the effects of technology. They frequently come to the fore in periods in which economies and societies are in the throes of sweeping transformation—like today. (Florida, 2013)

Florida contends that the rise of new technologies is an effect of (rather than the cause of) major societal changes. During periods of great change—such as when human economies progressed from a largely agricultural foundation to the industrial model that became dominant for much of the 20th Century—technological developments are stimulated: "innovation slows down during the highly speculative times leading up to great economic crises, only to surge forward as the crisis turns toward recovery" (Florida, 2013).

Florida speculates that our current period of socioeconomic evolution—from the industrial model of the 20th Century to the post-industrial, knowledge-driven, "creative economy" of today—is stimulating another surge of innovation, a surge with both dislocating consequences and tremendous potential for transformation. As this wave of technological innovation breaks across our global economy, Florida points towards the missed opportunities of past transformations and declares:

> We can't simply write off the tens of millions of workers who toil in dead-end service jobs, or the millions more who are unemployed and underemployed… we need a new social compact—a Creative Compact—that extends the advantages of our emergent knowledge and creative economy to a much broader range of workers. Every job must be "creatified"; we must harness the creativity of every single human being. (Florida, 2013)

It took more than half a century, a brutal worldwide Great Depression, and two World Wars (not to mention the societal and technological changes those wars produced), for Western economies to find an equilibrium after the Industrial Revolution. Other major world economies are still trying to manage this change in places like India and China. As we cross the boundary of another huge change in the way human economies operate, the question arises: can we learn from these past transformations and proactively manage present technological evolution in a way that benefits the broadest number of humans?

What does this have to do with Automation? Here we go!

Over thousands of years—a very short period of time evolutionarily speaking—humans created civilizations (and all its sub elements) and commerce, all to further the survival of the species. Civilization, in this framework, is all about improving survival; commerce is all about distributing resources. But how do we make the leap from the processes of human cultural evolution to the processes (like automation) that will help contemporary businesses evolve and thrive?

It's important here, more than ever, to bear in mind the analysis of economist and management expert, Nancy Adler: "Those individual businesses, firms, industries and whole societies that clearly understand the new rules of doing business in a world economy will prosper; those that do not will perish" (Adler, 2002). This sounds very much like evolution in response to environmental change, competition, and selective pressure to me. In fact, it is and always has been.

Above I used the term "unnatural competitive advantage," which could be read as the idea that concepts like money, commerce, and the economy somehow exist outside of the sphere of natural selection and evolution. I actually strongly argue the opposite. Although these human constructs are not strictly biological in nature (thus "unnatural"), they are simply non-biological expressions of the evolutionary process. We as humans—as creatures of an evolutionarily driven Earth—have no other option than to evolve in the manner that species have been evolving over millions of years. The specialized, non-biological adaptive responses humans have constructed to improve their chances of survival and growth are as susceptible to (perhaps even more susceptible to) evolutionary processes. In fact, it's the unforgiving nature of evolution that has made our civilizations and economic systems strong enough to endure across generations.

In the earliest days of the primordial soup, the fundamental basis for evolution was the trading of energy necessary to sustain particular chemical reactions. This process continued through the first multi-celled organisms, through every creature in the prehistoric oceans, to the dinosaurs, early mammals, and our own human ancestors. And it continues through every single human being alive today and into the future. The goal has always been (and remains) the efficient acquisition of energy. Humans have evolved the ability to conceive of advanced non-biological methods for acquiring energy (the system of commerce, etc.), but that simply makes us and our approach to survival the product of billions of years of evolutionary work, not something else that came about outside of the standard evolutionary processes.

Humans invented and utilized all sorts of tools, weapons, social structures, and complex behaviors, all for the same goal: gaining and retaining the energy needed to maintain chemical reactions in our bodies—descendants of those same reactions that started billions of years ago. We are just at another point on the spectrum of creatures with unique survival abilities shaped by evolution. Our evolutionary progress is plotted on an exponential curve, because this is the first time in history a species has been able to evolve via methods outside of biology alone, but the processes themselves remain mostly unchanged.

All the advanced systems and technologies we now create are extensions of that original, life creating, and sustaining process. I contend that as those technologies become more and more efficient over time, the ultimate result is increased automation. The shovel makes us more efficient at digging a hole in the dirt—presumably for some survival related effort—while automation is the ultimate expression of the use of machines to create efficient methods of survival. Technology advances (again, shaped by the evolutionary processes outlined above) to the point where production capacity is so great and so efficient that required human labor inputs fall to essentially zero. The automated future is inevitable because it is the natural extension of the evolutionary process, which has governed human direction and activity for millions of years.

So how exactly does that work?

It all begins with a process. Everything in the universe is a product of a process and a participant in a never ending series of changes that will continue through the end of time. I love processes and you do, too, even if you don't know it!

Processes exist, even when you did not deliberately put them in place. They're there. In fact, our greatest human skill is the ability to recognize process and/or the possibility of process, to deliberately adapt and execute a process of our choosing to ends dictated by our own desires. These processes take many forms: from mathematical formulas, to the tactical plans used to achieve successful business strategies and, of course, computer algorithms, which issue commands that are faithfully executed over and over without fail by machines. No idea can ever become reality without a repeatable process underpinning its core purpose.

And this brings us to the point where I make some big assumptions, because frankly this question is so big there is no other way to begin than with an assumption from which to work logically downward to see if we can connect the dots to achieve a logical proof (which of course I can).

From the discussion above, I propose that the meaning of life is to fulfill one's purpose. One's purpose, ultimately, is to survive. In order to survive, one must creatively envision a future reality in which they are alive (and preferably thriving). From here, one conceives and executes a process to manifest their vision into reality. If successful, reality bends to the will of the human. If unsuccessful the eventual result is death. Successfully achieving our vision of a future reality becomes our purpose by way of our need to stay alive. In order to stay alive, we must successfully create a reality in which our needs are met.

It's really as simple as that. See the future. Create the future. Live. Evolution has brought us to this point: where our very will determines our survival, not just our biology. Our bodies are our first technology. This was the first technology that allowed us to express our will, our vision for the future, and to make it a reality. Our body was our first technological tool from which ideas were forged into the world.

If staying alive is our number one purpose, technology is the functional expression of that purpose. A process is needed to achieve our purpose, technology is created to increase the efficiency of the execution of that process to achieve that purpose. Ever advancing technology has at last reached a point where once the human mind envisions a future reality and sets up a corresponding process, available technology can produce the output without additional human input.

Automation is born. We live.

1 THE FOUR STEPS OF APPLIED AUTOMATION

Now that I've explained *why* I believe that automation is the inevitable next evolutionary step in business processes, I'd like to spend some time expanding on how we've reached this point and what that history reveals about future evolution. How did humans move from utilizing strictly biological processes to using complex technological processes? Does the logic of that progression continue to drive change? And if these evolutions of process are indeed inevitable, can understanding the mechanisms of past change help us to prepare for the next leap forward?

As you have no doubt gathered, I believe automation is the holy grail of building a business. As we continue to evolve as a species, we will see more of the everyday processes that make our lives easier start to evolve towards automation. Today, some business processes are completely automated while others continue to rely on human intervention due to limitations in technology (or resistance to change, as we'll explore in later chapters).

This chapter will focus on describing the four stages of applied automation, which occur as technology progresses toward the point where it becomes fully automated. The four stages describe (in simplified terms) the evolution of technology across time. This process, while much faster than biological evolution, has historically taken substantial time to unfold. Today, given advances in technology to this point, it's happening so rapidly that we can track this evolution on a yearly basis! I'm going to start off with a simple metaphor that makes it easier to see all four stages of applied automation.

The Lighthouse Metaphor

I like to use this example as an opener to help explain applied automation. This is something that everyone can easily relate to and intuitively grasp without the need to dig through complicated theory or confusing business lingo.

Lighthouses, in one form or another, have existed for thousands of years and in that time have evolved from completely manual to fully automated processes. Lighthouse development is well understood and provides a great illustration of the evolution of technology that can be seen very easily. Are you ready to get started?

In the earliest days (one imagines), there was a man who lived near the ocean on a hill. Observing passing ships, he noticed a reoccurring problem: ships venturing too close to land often ran aground on dangerous, hidden rocks below the surface, always in the same spot. After seeing so much misfortune, the man decided to take it upon himself to help. Therefore, every time he happened to see a boat heading for that crash spot, he would start shouting out to them to "avoid the rocks!" This actually represents stage one of the applied automation process, which is entirely manual. The man noticed a problem and wanted to fix it. He developed a process, which was to shout out and warn incoming boats. The only problem was that his method was not efficient. What happened if he didn't notice an approaching boat in time? What if he was sleeping or otherwise engaged? His process of warning ships was completely manual, with no contingencies for such variables.

Eventually, other people noticed the same trend as the man on the hill and used humanity's most powerful tool–our imagination–to evolve this process. That step in evolution came with the idea to build lighthouses. They built a tower and put a light at the top of it to warn boats to steer away. Isn't that much more efficient than a man shouting out to incoming boats?

Clearly, it is. But they still had to find a lighthouse keeper to manage each lighthouse. Even though they had effective technology in place, it was still dependent on human labor to maintain the lighthouse. If the lighthouse keeper didn't put oil in the lamps every day, the light would go out leaving ships unprotected.

So what has happened here so far? Two things. The first occurred when our hero initiated the original manual process. He analyzed the ongoing problem of shipwrecks and formed a set process that determined what needed to happen to solve that problem. Even though he was doing it inefficiently, his initial process made it possible to evolve. The second thing that happened was that human creativity evolved that process to its next step: a more efficient system. That evolutionary step replaced the guy shouting to incoming ships with a lighthouse that shined a warning light. This step was much more efficient, but still dependent on human input in the form of the lighthouse keeper maintaining the fire. The lighthouse was stage two of the applied automation process of warning boats of danger.

Lighthouse evolution took another step forward when the Fresnel Lens was invented. It was a great jump in technology since the light could be seen from further away, but it was still just as dependent on human labor. If the lighthouse keeper didn't do his job, then the entire system would shut down. So: this technology simply *improved* upon stage two of the applied automation process by allowing a single lighthouse to service more boats, but it was not quite enough to evolve out of stage two.

I'll quickly note here that this is the stage where most of humanity exists at this point in time. We have the technology to make our lives much easier (smartphones, tablets, etc.), but this technology is still dependent on us to operate it. I'll discuss this in more detail at the end of this chapter, but I just want you to make a quick note in your head.

With the development of the Fresnel Lens, humans had extended the range of the lighthouse and made it much more efficient, but since it still required a lighthouse keeper to maintain it, a significant potential bottleneck in the process remained. Until this potential breakdown in the system was corrected, the risk of shipwreck was still present.

In business, we are constantly on the lookout for limiting factors, like the lighthouse keeper. We refer to these limiting factors as bottlenecks. A business can have unlimited capacity, but bottlenecks will limit its potential. Take a moment here to consider the bottlenecks in your own business and write them down as you think of new ones.

Okay, by now the lighthouse process was close to evolving out of stage two, but was still missing some technology: namely electric light and solar panels. Once that technology became available, it allowed the lighthouse process to evolve to stage three. Slap on some solar panels, throw in an electric light with a sensor to tell it to shut off during the day, and guess what? We no longer need the lighthouse keeper. We have evolved to the point of removing the bottleneck. Our lighthouse has become fully automated.

Now, that same process is much more reliable than when it was dependent on that lighthouse keeper. Think about the implications: what would have happened if the lighthouse keeper had gotten sick or overslept? The entire process would have been compromised. Moving to stage three of automation, that potential for error has been eliminated. Our lighthouse is self-governing, adaptable to changes in conditions (such as shortened daylight hours, etc.) and, most importantly, completely reliable (with proper periodic maintenance, of course).

Transcending stages one and two required additional technology and the removal of human input as a bottleneck. But aren't there four stages of applied automation? How can we improve on our stage three development from here?

The lighthouse process evolved to <u>stage four</u> when we stopped relying solely on a lighthouse to protect ships. We now have ship-board GPS systems that are extremely reliable and powerful, designed to provide accurate information on any potential dangers. Lighthouses are still in place simply for emergencies, but they are no longer necessary.

The GPS system is a network of different processes that rely on the same (or similar) outputs. Since there is a very clear definition involved (coordinates), it can be used to completely automate a ton of different processes that rely on coordinates. I hope I haven't confused anyone with this last step. It will be completely clear once this chapter is finished.

Either way, the lighthouse metaphor provides a clear example of the four stages of applied automation.

The Four Stages of Applied Automation

Let's review the four stages once more in broader terms, now that you have the lighthouse metaphor to fall back on for reference.

Stage 1: All Manual

Everything starts out manually. In the lighthouse process, it started with the man shouting to boats. He was experimenting with his physical senses to accomplish a goal. There's a lot of potential for trial and error here. We may try several possible solutions and when they fail, we use our imagination to find new possibilities. It's our imagination that sets us apart from all other species on this planet. Every other living thing on this planet exists in stage one. It's our ability to imagine new methods of doing things that allows us to move past this stage.

Stage 2: Automation Generation Zero

Stage two is basically *manual with improved capability*. Everything is still done manually but we are using more sophisticated tools to perform the processes. These tools do not function without human input. Think about the beginning of the lighthouse. It automatically warned boats to avoid specific areas, but relied on human input to keep the light burning, just like your smartphone requires you to push buttons to tell it exactly what to do. It doesn't really do anything by itself: it's a tool. Without someone operating it, it doesn't really do anything on its own (except maybe install updates).

Again, this stage encompasses the overwhelming majority our technological evolution at this point.

Stage 3: *Automation Generation One*

Stage three is where the world is quickly headed: we are beginning to see the connection of processes. Suddenly, the process evolves from a mere tool, to multiple individuals linking together to accomplish the same goal. Processes start to connect directly to each other, allowing technology to make decisions and run the entire operation. All that's required is that the processes involved are all set to achieve a specifically designed goal.

There are some major advantages to this step, including:

- Limiting labor
- Limits human intervention
- Cuts costs
- Increases efficiency
- Increases accuracy

Stage 4: *Automation Generation Two*

The fourth stage of automation is still on the far horizon for most business processes; I can see stage four becoming realistic and widespread in the next thirty or so years. As automation becomes the territory upon which our commercial battles are fought, the same selective pressures we encountered when looking at biological evolution will begin to declare winners and losers in the technology realm, too. Businesses will become fully automated or face extinction: the business environment will require it. You will also start seeing automated processes being implemented between businesses. A process that runs within one business will suddenly jump into the next business directly.

You don't really see this a lot in businesses now, but there is one example where you can see this stage in our present marketplace: in stock markets. As any experienced investor knows, people don't manually trade stocks anymore. What happens instead is this:

Person A inputs a set of variables into their trading software. Let's say that they are looking to buy a specific stock for a specific price. They can input this information into their software and then that computer will go out and start looking for a seller with matching criteria. On the other end of this burgeoning transaction, Person B will have told their software that they are looking to sell that same stock at a price that falls in the price range of Person A's criteria.

At this point, the process runs itself. These two computers will search potential trading partners until they find each other and automatically make the deal. The transaction itself is completely automated. This happens many millions of times every day in stock markets around the world, albeit on a much more advanced level. Traders input their instructions and the machines complete them without additional input. This can happen because:

1. The process of trading has been established.
2. Both traders have very specific goals that align.

Now, imagine this same sort of process happening across manufacturing operations. Let's use the trucking process as an example. A company is looking to have a product shipped. They create a shipping budget that sets their pricing parameters and they send out an automated request to all of the trucking companies servicing their area. These trucking companies, in turn, have a system in place that can automatically receive requests, process their details, and then reply with a bid.

This process is being done right now, of course, but instead of computers reaching out to other computers, people are calling other people (or sending emails). However, I can see this process becoming completely automated in the near future, since there are so many common variables in place.

A Practical Example

Let's take a look at a common concept in business called "*lights out.*" This concept describes a company (or a component of a company, such as a factory) that doesn't need any employees to operate. A "lights out" factory could simply turn off the lights and let the machines run in the dark.

Presently, "lights out" factories are still the exception. But we are quickly approaching this lights out era in business, or at least developing the technologies to make it widely possible. What's holding us back? Very little, really, other than a society that is stubbornly affixed in stage two of the applied automation process. Yes, we have a lot of awesome tools—which save time and greatly improve productivity—but they still require our input to function.

Now let's turn to a practical example of stage three automation. Let's say that you own a business that's developed to the point where you can fully automate it. When an order arrives, the entire process can be automated, all the way to the point of an invoice being sent out and payment automatically remitted by the customer. Heck, even the process of getting paid can be easily automated! In this example, the following process would happen:

1. An automated system receives an order and processes it.
2. That order is fulfilled.
3. That order is automatically delivered.

4. An invoice is sent.
5. Payment is automatically received and processed.

In this example, there would be no human interaction involved from the time an order is placed until payment is confirmed.

Now let's assume we put this sophisticated system to use. There is still one key element missing—the sales team. In today's world, we still need a sales team that markets our business to the world. This is what is truly holding us back from a fully lights out enterprise. A business has to make a connection with a potential customer and a human sales team is still the tool we use most often to accomplish that goal. There is no standard protocol for that sales effort. The art of selling requires creativity, so we still need people for this important step. We have the technology right now to fully automate a lot of businesses, but that human interaction remains critical to making an initial connection in most cases.

If we ever do reach stage four of applied automation, we will not even need that sales interaction. But this will only happen when we establish a specific protocol for intra-business communication. Once that happens, internal processes will be able to communicate, reaching out to other businesses directly and automatically, and completing sales independently in much the same way they do in today's stock market.

Where Our Technological Evolution Currently Exists

We are still comfortably rooted in the *manual with improved capabilities* stage of applied automation. Some businesses are moving in the direction of full automation, but human intervention is still required in most cases.

We have computers, smartphones, and many other technological marvels that make our lives much easier. But even so, this technology still requires our input to do the remarkable things they do. This makes our lives easier and more productive, but we are still required to attend to our technology and guide its functions. Technology has automated certain portions of the business process, but for the most part it has only allowed a few things to truly become truly "lights out." *One or two things out of a thousand processes are fully automated.*

We're doing the same tasks that we've always done—and we're doing them better—but our constant vigilance is required. As we'll see, that vigilance can be difficult to sustain.

The "Problem" With People

I use the word "problem" very lightly here. People are not really a problem, but they do bottleneck businesses; often simply by being human!

The first common "problem" is simple boredom. I suffer from this problem myself. I get bored quite easily and what happens when I get bored? I weaken the system. Maybe I'm not doing as much as I should be or the monotony of what I should be doing causes my attention to wander. Although these lapses don't shut down the productivity of a business completely, it does slow it down. Even with the advanced tools that we use at this stage, the problem of boredom is pervasive. And because even the most focused human minds tend to tire or become bored with repetitive tasks, it's a problem that isn't going away anytime soon.

Despite this (or perhaps because of it), I believe we're quickly approaching stage three. More businesses are truly becoming automated to the extent that current technology will allow. Businesses want to utilize less people. Or rather, businesses want to use people only for those tasks that require human talents. Automation makes this possible.

What Happens When a System Becomes Fully Automated?

When you go to a fully automated system, there are going to be two main issues. One is that people will need to maintain the system. However, this will involve far fewer employees than the process itself, so each employee's productivity goes through the roof. Still, you'll need at least one person there.

The second issue is that you'll need a machine to perform the process itself. Machines do break down, so there will be downtime associated with a fully automated system. This downtime might happen once a year or once a decade, but it will happen. However, this is not really the problem it seems. In fact, it's actually a good thing. Here's why.

When people are performing these processes, you are still getting downtime, but it's not measurable. It's what I call the *"death by a thousand small cuts"* scenario. Each "cut"—in this case human-caused downtime—is too small to measure individually. But cumulatively, these cuts will slowly bleed energy and momentum from the system. Left uncorrected, your system faces collapse, or in the language of our evolutionary framework: extinction.

Conversely, with a fully automated system, it's either working at 100% capacity or not working at all. There are no "small cuts" that will bleed the process.

What Are These "Small Cuts?"

People need vacations, get sick, or as I mentioned earlier, get bored. This results in unpredictable fluctuations in their productivity: sometimes they perform great, while at other times they are not operating at peak efficiency.

Take the example of an employee returning to work after a vacation. They require time to get back into the groove, to go through pages of emails, to catch up on new business developments and get back into the routine of the work week. This ramp up period—especially when multiplied across many employees shifting into and out of their normal routines—is a small cut that can gradually bleed a company. Worst of all, each cut is too small to easily detect or correct. Individually the damage is insignificant, but cumulatively it can spell disaster.

This ramp up period—like many of the all too human imperfections that slow business processes—is completely eliminated with a fully automated system. When your automated system isn't working, you will know it. You will have an estimated repair time, so the cost of downtime becomes much more quantifiable. By eliminating tiny inconsistencies, you are increasing cumulative efficiency of the process (instead of the other way around).

And businesses are noticing. People aren't completely satisfied with our enhanced tools. They are realizing that tools that require human input are only as useful as their human users make them. The option of automation—while still somewhat limited by available technology—can eliminate productivity flux and free humans to pursue tasks for which they are exclusively adapted.

In the next chapter, we'll take a look at the option of business automation in more detail: how do businesses approach this potential change, analyze their needs, and take the great leap into an automated future?

2 THE PROCESSES OF AUTOMATION

This chapter is going to take you through some examples of different processes that companies can automate. We'll walk through several examples of the process and discuss some major ideas and themes that will be important for understanding later sections of this book.

Much of the time, decision makers in the companies I work with don't even realize that some of the processes they are performing manually can be automated. After all, these are things they've been doing manually for years. They have acquired an arsenal of cool tools that seem powerful, so they tend to believe those tools are the maximum amount of technological enhancement possible at this point. They don't even realize that there is actually a better way–a more automated way–to accomplish that same business task.

Status Quo Bias

The phenomenon described above is generally caused by a pernicious, but often difficult-to-avoid pitfall called "status quo bias." Status quo bias is defined as "the irrational preference for one option over another simply because it preserves the status quo" (Bostrum & Ord, 2007). In other words, humans have a strong tendency to stick with current ways of operating, even when other options make more sense rationally. Our evolutionary history is partly to blame for this: when humans were struggling to find the resources simply to avoid starvation, the proverbial "bird in the hand" really was worth the "two in the bush," since there was no way of knowing how much additional energy would be used pursuing the unknown. We're wired to conserve resources and prevent loss. These days, we often rationalize this bias by (usually falsely) convincing ourselves that our situation actually is the best option currently available: we have all the tools, the software packages, the indispensable personnel! How can the system be improved upon? In my business, this often leads to resistance to automation or a "small bore" approach, where only a limited number of processes are believed to be ripe for automation.

I actually see this a lot. I visit a company and observe that they are focused in on a small process that they feel is important: the real problem. They think it can be automated, so they are prepared to take that mental leap in one particular area. Then I come in and look at their idea only to conclude that there are many other processes ready to be automated that they have not even recognized yet. These are always a lot of fun to point out.

Those are areas that we are going to discuss in this chapter. I'm going to walk you through several examples of processes that are not easily recognizable by many businesses. These areas are where they should be focusing from an automation standpoint. There is a huge difference between individual tools that increase efficiency and total automation of a process. Tools require a person to make them work, while automation happens on its own based on the system that's put in place.

Why is Automation Associated with Robots?

"Robot" is a word that always comes up in discussions that have to do with automation. That's partly because the word automation causes previous generations to instantly envision those robots that work in car factories. Around for decades, they start to visualize one of those mechanized robots with arms and legs.

However, when we discuss automation, we are not talking about these mechanical robots. We are talking about a complex series of processes that are executed automatically free of human labor.

What is a Robot?

At its root, a robot is a machine capable of carrying out a complex series of actions automatically. When we build a robot, it doesn't necessarily have arms and legs because having arms and legs are not necessary to the job it's going to perform. If a robot is tasked with processing and distributing data in a particular way, then it doesn't need arms or legs. That makes the whole science fiction depiction of robots inaccurate. Sure, we will probably build robots like those at some point, but the vast majority are going to be the ones that you don't actually see. Their job is to process tasks behind the scenes. They have other types of appendages besides arms and legs.

I will use a concept that most people are familiar with to explain this:

automated homes, or "smart homes" as they are often marketed. The process of home automation relies on many sensors being installed throughout the house: sensors that turn lights on/off or sensors that measure the temperature in a hot tub and adjust it accordingly. To keep it simple, let's just use light sensors for this example. If you were to install these throughout the home so that lights automatically turn off when the room is empty and turn on when someone enters, then you will have a robotic house. You are not pushing buttons to control each decision. All you have to do is set a variety of parameters and the house will automatically respond to those parameters, reliably, until the parameters are changed.

In this robotic house, a complex series of actions happen based on preset parameters. This is closer to the type of automation we're discussing in this book. It doesn't necessarily have anything to do with mechanization.

Business Automation

In business automation, the goal of any initiative should always be that the system controls the business processes. I like to use McDonalds as an example of how this works since they actually use a process (which most of us are familiar with, even if we never worked at McDonalds!) that could be almost completely automated. Let's look at this example more closely.

The McDonalds Robot

For the purposes of this example, it's useful to think of the entire restaurant as a robot. Here's how it performs its job:

1. Customers place an order and an employee enters that order into the system (input).
2. The order is sent to a monitor in the kitchen where the employee in back can see it.
3. The employee in the back makes the order.
4. The customer gets their order.

Those are the basics, but let's dive in just a little deeper.

When you walk into McDonalds and place your order, the employee behind the counter inputs your order into the system. That person has been trained to push the buttons on the screen that places your order into the system. Now, when the company first implemented this system, they had to have that person behind the counter pushing the buttons. At this point, however, that person is not really necessary. I don't know why McDonalds has not gone to an automated system, where the customer pushes the button themselves. I feel it's probably the expense of implementing a company-wide system which would be very expensive. Or perhaps it's a combination of expense and status quo bias. In any event, as the company's labor costs rise (as is probable with current efforts to raise the minimum wage), the expense of automation vs. status quo labor will become more attractive.

But let's return to our example and continue following the chain of processes that produce your hamburger. So far, you have the employee behind the counter who inputs your order and then that order gets processed and sent to the kitchen. The cook in the back has the arms required to make a burger in a manufacturing style. By manufacturing style, I mean that they have a set process relating to what items they stack and in what order they stack those items. So the cook looks up at the monitor to see what he's supposed to make and then looks down to make the necessary items. And that very specific process is repeated throughout the day, each time an order is placed. All orders come through the system and are fulfilled in a specified manufacturing style.

Now let's approach this example from a larger business perspective. The goal of an automated system is to remove all room for error and creativity from the business process. By the time an automated process is put in place, the creative part of the job is long completed. Creativity comes when designing these processes–bringing something new into a business is creative. Once the process is in place, there really should be no creativity involved. It should be a repetitive action at that point.

Once it becomes a repetitive action, then sometimes people are required because of technological limitations. McDonalds cannot buy a burger building robot at this time (or at least enough burger building robots to staff all their restaurants), so they still need that employee working in the back to monitor and fill orders. Instead of automating, McDonalds systematized the burger building process to leave less room for error. Since the process is so precise, there is no way they could forget the order (since it's displayed on a screen). And the process for making the food is essentially set in stone. There is no room for making decisions that aren't predefined by whoever designed the business process.

This same goal should be in mind when building any automated business process or with any system that's used in a business.

It Has to be a Company-Wide Initiative

This is probably the reason why McDonalds has not implemented fully automated systems at this time. The expense of automating all of their locations would likely cost billions.

If it's not company-wide then you end up with increased efficiency in a particular area, yet you have to link those areas with the rest using some form of software. This link becomes dependent on people in the middle to keep the transactions rolling. So what happens if this middle person gets sick? Everything comes to a screeching halt! All of the efficiencies that are in place in one area are then completely undermined because of one person getting sick.

Hot Swapping

Hot swapping refers to the ability to switch components (in this example employees) in and out of a system without shutting the system down to do it. McDonalds serves as a great example of hot swapping. They have a system in place that is so well designed that they can put anyone in any position to keep it running. You want the machines to do as much as possible on their own. For the areas where the process needs a person, then you want to be able to put anyone in that slot without a lot of specific training. Anyone should be able to give the machine what it needs at that particular moment in the process.

This falls under the design of the process. At this point, it has very little to do with technology and more to do with who is designing the system and how they are approaching the different challenges in a particular business. The designers simply should not allow these pockets of people dependency.

A Practical Example

Now let's look at another example of something that my company ran into. We were inspecting a business automation project for a ground transportation company working with the airline industry, moving pilots back and forth between hotels and airports. This wasn't a small company either: between all of their locations they probably made a thousand trips every day.

We designed a system for them that would basically automate the entire process with the exception of the driver. At this point in our technological development, we do not have automated drivers so the overall process still required a human to cover that position.

The key to automation came in the form of GPS. Both the customer and driver had GPS access so that they could find each other. Since they could easily meet up, a dispatcher was not a requirement for the process. The automated system was linked directly to the airline so that the system could see flight schedules in order to determine when flights were running late. The system automatically updated the driver schedule and kept everybody moving where they needed to be based on real time information.

Up to this point, the company was doing everything manually. Dispatchers were manually handling a thousand trips on a daily basis across the entire country. They were calling drivers, setting up meeting points, and doing all of those processes manually.

When we presented our automated system to this company, they decided that they really liked it and wanted to move forward—but with one critical condition. They only wanted to implement the automated system on the customer experience side and not on the back end processes. They told us that their "primary concern is the customer experience, so that's the area we want to focus on. Let the back end processes stay manual."

Here's the thing: it's just not practical to build a system like that. When you set up all of these automated capabilities on the customer side, it has to be linked to the back end stuff for it to work. It will need to be connected with the scheduling system as well as the management system. All of this back end stuff has to be automated in order to truly make the process easier on the customer. The back end machine must be in place so that the front end system will work correctly. To use a slightly silly metaphor: imagine a whale that has suddenly acquired the ability to eat and gain energy efficiently from soy beans. It's certainly a great leap forward for the whale's dietary options. But since soy beans don't grow in the ocean and whales aren't able to travel on land, this potentially important change to the system can't be fully exploited (until the whale grows legs). An automated front end that links to an inefficient, manual back-end ends up in a similar place: with potentially great advances hobbled by half measures.

The company's Head of Dispatch explained his idea: he proposed having all of the customer facing processes automated, so that when they came through the system, they would come to a dispatcher who would enter them into the back end part of the process. That would allow them to keep the back end processes manual while allowing the customer experience side to be fully automated. In other words, all of these requests would come in and then a dispatcher would sit there in the middle of the process, look at the old manual system, and respond accordingly.

I'm not saying this idea wouldn't work (though I doubt it would work well). From the customer's perspective, it might appear as if everything is connected and working perfectly. However, there would still be a person in the middle responsible for information flow. It was a good solution for what management wanted to do. This type of system would technically work, but it doesn't take a rocket scientist to see the numerous issues that would come up with this setup. I mean, we are talking about building an automated system on the front end while having to connect to a manual system. To be perfectly honest, it would be a disaster!

I actually made a joke about it in the meeting when this idea was presented. It reminded me of one of those old Warner Brothers cartoons where they build this big robot to perform tasks. However, inside of this robot's belly was a hamster on a wheel making it go. That's a fair analogy to how a system that's half automated/half manual would work. Those are the kinds of situations that you really want to avoid, where your newly automated system still depends on timely human input to function. What happens when the person connecting the systems gets sick? What happens if that person makes a mistake? The whole system fails.

The moral of that story is this: automate everything, whenever possible. Trying to automate half of a process and connecting it to a manual system can only lead to disaster.

In the end, the company was unable to envision the benefits of a fully automated system and backed out of the deal in the final stages of planning, electing to continue with the cost and stress of the status quo.

The Right Tool for the Job

Ultimately, the lesson of this example—and a concept that we will return to throughout this book—is a simple one: use the right tool for the job. As the cognitive scientist Douglas Hofstadter put it more than thirty years ago: " if someone says a task is 'mechanical,' it doesn't mean that people are incapable of doing the task; it implies, though, that only a machine could do it over and over without ever complaining or feeling bored" (Hofstadter, 1979). An automated business process like the one described above takes advantage of this truism, utilizing the tools of automation to accomplish the un-creative task of mechanically matching the transportation needs of individuals to evolving flight schedule data in the most efficient way possible. Human labor, with its unique creative capacity, is thus free to step outside this formal system of transport matchmaking and focus their analytical and creative faculties on conceiving further improvements to the system or even new areas of potential business expansion. To again quote Hofstadter, "Computers by their very nature are the most inflexible, desireless, rule-following of beasts." At least for now, humans alone have the ability to analyze processes, measure their output against human desires, and create new rules and processes to meet the scope of those desires.

The Concept of Marketing Automation

Marketing automation is a concept that you've probably heard a lot about. It's really caught on in the last few years. Marketing automation usually revolves around social media. Social media is easy to automate since there are a wide variety of programs out there devoted to this exact purpose. Every social media site follows a standard protocol, which makes the process simple to automate. You can easily set up systems that automatically post the same thing to all of your social media accounts or schedule many postings in advance.

Marketing automation has become my generation's default image of what automation means, just as the previous generation looked at automation as mechanized robots working in factories.

Marketing automation is a great idea, but I tend to look at it as just automating a particular part of the business. It's helpful, but it's still just a productivity tool. You see, marketing automation doesn't really help with marketing: it just speeds it up. A person is still required to put that strategy together. This will always be the case since marketing requires creativity.

The problem is that many people believe that marketing automation is the limitation of their business automation. Getting our clients to think about automation outside of the marketing realm can be challenging.

Starting the Journey to Automation

Here's the meat of what a business can automate. Look at all of the back end processes including:

- The workflow processes
- The sales processes
- Some of the marketing processes (like social media)

Let's look at another example, this time on the financial side of business. Many companies use QuickBooks. It's a great accounting system, perfect for small and medium sized businesses. In fact, even some large companies are use QuickBooks Enterprise Edition. As useful as it is, QuickBooks is just an accounting system at the end of the day. It is tasked with keeping track of finances. It still requires input from people, so if the numbers are put in wrong, then it's not going to do the job correctly. That makes QuickBooks just a productivity tool.

When we first look for areas to automate in a business, we generally start on the financial side. Is there an accounting department involved? How are invoices being generated?

A lot of times, we see companies still generating invoices manually, especially when they are using QuickBooks. Each time they generate an invoice, they have to call the head of production and ask them for the details of a particular deal.

Other times, the person generating the invoice can find the information in a manual system, so they click around to get it. They get the results they need eventually, but the process takes time and effort. My main point here is that a person is gathering all of this information (sometimes retrieving it from the system, sometimes by asking another person for details) and then manually entering it into QuickBooks to generate an invoice. If a person can find this information, it means that the information is readily available. And if it requires a phone call, then that company is way behind the times already.

Often when I go into a company, one of the first things I notice are divergent systems. The shipping department might have one system, while the production department has a completely different system. And of course, these independent systems don't often communicate well with each other. Therefore, when a bridge needs to be built between divergent systems—for billing purposes or other reasons—it usually takes people to gather and organize this information together into a meaningful form.

In cases like this, the goal of automation is to create processes that pull that information together by linking those systems through a master automation system. By doing so, the creation of an invoice—or any other commonly used report or information query—can be fully automated.

Another thing that we see a lot of is the use of Excel spreadsheets in the business management process. Surprisingly, it's the big companies that are notorious for this. They manage their workflow using these complicated spreadsheets that they have created. There are some obvious issues with this strategy:

1. The process is time consuming.
2. Data input is prone to error. When a person makes a mistake, there are no checks and balances to catch those mistakes.
3. Using those spreadsheets requires training.

In cases like this, the system doesn't verify anything: it just keeps records, accurate or not. If the records are created with errors, there is no good way of catching it. Honestly, spreadsheets remind me of those old banking ledgers. That's really all a spreadsheet is. The only legitimate difference is that a spreadsheet is electronically stored. If you find your business is dependent on Excel spreadsheets, then you're not that far ahead of those old banking ledgers where everything was written by hand.

What I normally see are huge amounts of spreadsheets being used as part of the business process. When I see this, it automatically tells me that there is plenty of room for improvement.

Inputs and Their Role

For an automated system to function, it has to have sensory input from the outside world. Most of the time, that input is in the form of keyboard, mouse, and a person who gives the system what it wants. The easiest example of this is a scale.

Let's assume that a business uses a scale to weigh pallets. The system will ask how much a pallet weighs. So the input would be the product on the pallet and the weight on the scale. Those are the inputs that the system is looking for.

Of course, in my world of automation the scale would be digital and linked directly into the system. A worker would only need to put the pallet on the scale and scan it with a barcode reader. The system would then enter the weight automatically, assigning that information to the pallet's specific barcode for easy reference. Instead of manually entering the product and weight into a system—a more inefficient process, with more room for error—the worker would simply scan to give the system all the information it needs.

Today's technology requires a person to at least walk around and barcode things to tell the system what it is. Then they would have to put it on the scale because we don't have the mechanized robots yet to do that. This is an example of how sensory input functions.

When you have an automated process, all of those input devices are what gives the system its ability to know what's going on in the world. The idea is to let those devices do more and more of the work. Instead of a person typing in the actual weight from a manual scale, let's get a digital scale and have it hooked directly into the system. That way no one has to type anything in. It's faster and more accurate.

Example: Excel Spreadsheets Gone Wrong

By now, you might be asking yourself why a large company would still be dependent on Excel spreadsheets. When this happens, it's a clear sign that the business has grown quickly. Small companies can effectively use spreadsheets to keep track of everything. But when they grow quickly, then these systems become a problem rather than a solution. Companies tend to focus on their growth and don't monitor or update their processes as often as they should. The greater the growth, the more responsibility these imperfect Excel-based systems take on. Eventually they reach a breaking point.

Let me walk you through an example of an Excel spreadsheet systems gone wrong. I received a call from a food distribution center. For four years, this company had grown at the rapid pace of 50% every year. They were growing so rapidly, they needed administrative help, so after about two years they hired a Chief Operating Officer. This lady is brilliant and instantly recognized that the system they had in place when she started was–well, the problem was that there was no true system in place. They were still tracking everything on paper.

The COO realized that this was unsustainable, so she created an Excel spreadsheet based system. And at that point in the company's growth, it solved most of their problems. Moving from paper to spreadsheets was a giant step in the right direction.

Then two years later, she called us because she knew that the system that she had built had been outgrown. She explained to us that the spreadsheet system they had in place was making errors nearly 50% of the time. So basically, the whole system was always 50% wrong!

The reason was simple to track down: their spreadsheets were all linked together. They had one master spreadsheet in place that served as their master scheduling system, so when they got an order, that master spreadsheet was where the details of those orders were entered. Working from the master spreadsheet, they knew what order they had to fill.

The problem was that there were six different departments who would all get a copy of that master spreadsheet. It was updated daily and then sent out to all six departments. Each of these six departments had their own complex spreadsheet. So these departments would use the master spreadsheet to update their own complex spreadsheet. This allowed them to manually create their own workflow spreadsheets. Keep in mind that all six departments were doing this on a daily basis.

In an ideal world where everything goes right, I suppose that system could work. But we don't live in an ideal world. What ended up happening 50% of the time was that somewhere within those seven spreadsheets, someone would make a mistake or something would get misplaced. Suddenly, the spreadsheets didn't match up. Obviously, they had a problem trying to coordinate on the other end when these spreadsheets came together for the actual deliveries. Everyone would then spend valuable time trying to figure out what went wrong.

Half of their administrative staff in all of these areas were dedicated to trying to keep this system up and running accurately. So the cost involved in that system grew quite high. When we estimated this cost, it came to hundreds of thousands per year. This is not even taking into account the negative value of making so many mistakes to the customer experience. That's always an important factor to keep in mind when trying to value an automated system.

That is an example of an Excel spreadsheet system gone wrong. To make matters worse, all of their accounting was being done in QuickBooks, which wasn't connected to anything. So at the end of the day, they had a similar invoicing situation that I described earlier where they were slowly piecing information together from multiple departments.

Victims of Their Own Success

I know that this scenario sounds crazy. To some, it might even sound impossible. How can a large, successful company end up in this sort of situation? Actually, I see this sort of thing all of the time with very successful companies.

In cases like this one, the company has become a victim of its own success. Their rapid growth quickly outpaced their ability to create manageable processes. The system that worked at one level couldn't cope with the volume of information generated when the business grew. Can you imagine how disheartening that would feel? Your business is growing by leaps and bounds, your profits are fantastic, but instead of enjoying this success, you instead find yourself constantly repairing breakdowns in the process and hunting for the sources of errors. When a company like this calls, I know that management will often be more open to automation possibilities. It offers them the possibility of actually enjoying their successes instead of running to keep up with the growth of the company.

Currently, we are talking with a company that brings in $2 billion in revenue annually. In order to keep tabs on all that business volume, the company does regular audits across their 200 different locations. Their current system gives these 200 locations autonomy over how they perform these audits. So: 200 locations are performing this process, each with their own system, and then kicking their data back to the corporate level in whatever format they choose to use. The corporate level then has to parse all of this irregularly formatted data to get the information they need.

Let's do the math here: we have 200 locations and each location has 200 people they have to audit. That means that this $2 billion a year business is managing 40,000 audits manually! They are a very successful company, yet have still fallen prey to this issue. It's a symptom of uneven growth between business volume and tracking processes. Companies are focused on growing and being successful, but without proper attention their operational sides start to fall apart.

That's where the true value of automation comes in. Businesses want to focus on growth, success, and creativity. With an automated system in place, you can do just that. You can keep feeding more business into it, expanding volume is no problem. The system doesn't care if you add 100 transactions or 100 million transactions. With an automated system, you don't have to keep adding people to keep up. You don't have to keep training new people. Automated systems have essentially unlimited capability: if you need more capacity, you can simply add another server and keep rolling along.

Once you have those processes automated, you can focus your efforts on what got you into business in the first place. The operational side is necessary for business, but it doesn't drive growth. When you automate, you can devote less resources to this operational side and focus solely on growth.

3 EXCUSES COMPANIES MAKE TO NOT AUTOMATE THEIR PROCESSES

I have been in this business for over a decade. In that time, I've met with a lot of entrepreneurs and executives interested in automating their businesses. Some of the time these meetings are just the beginning steps of the process. I usually review their business and then present them with ideas about processes that can be automated. Some of them follow this plan through all the way to the end while others always find an excuse to somehow delay or avoid the big leap.

This chapter is going to reveal some of the most common excuses that companies use to avoid automating. Most of the time these decision makers are simply afraid. As I mentioned earlier, humans are hard-wired by evolutionary history to conserve resources and avoid losing them. This leads to the common human anxieties associated with change. Combine this natural aversion to complex automation ideas that they may feel unsure about, fears about the difficulties of implementation, and a host of other large and small concerns, and you can generate a fairly consuming dose of fear. And that fear—or rather the desire to avoid facing it—generates the excuses we'll investigate below.

These excuses normally come up as we are actually walking executives through the steps required to value what the automated solution will do for their company. On paper, automation is a "no-brainer" and makes logical sense. That's not what drives people away. It's the commitment that it takes to implement an automated system that drives that fear. Why? They are either uninformed or lack the support required to make these implementations easy. Naturally, they are afraid. It's at this point when they start finding reasons that give them an excuse to not go forward with an automated system.

Before I walk you through the most common excuses though, I want to take a moment to look at the difference between off the shelf products versus custom automated systems. Misconceptions about this subject often play a key role in the reasoning behind decisions to not go with an automated system. I believe that when you look at these two potential solutions rationally, with all the facts, the decision is much clearer than it may originally appear.

Off the Shelf Products vs. Custom Automation Systems

Since the rise of personal computing, software has become a mainstream part of everyone's life. When we run into a problem that our computers can't easily tackle, we're used to seeking an off the shelf product to meet that particular need. Developers build these products, copy them, and then sell them to millions of people with all manner of varied and complicated needs. This "one size fits all" solution has been the business model across the entire software industry for the past two decades.

The whole idea of custom implementation, on the other hand, has generally been reserved for computer techs who build stuff simply because they can or the high end corporations who can devote the resources to this need. Everyone else has relied on off the shelf products because the cost of custom products was thought prohibitive. Off the shelf software seemed like the most cost effective thing to do. And for the companies who develop software, it's clearly the most profitable thing to do. They have no motivation to tinker with this highly successful model.

The problem with off the shelf products from an automation perspective can be summed up in five words: they are not specific enough. These products are productivity tools, not automation tools. They are designed to create efficiency in a particular part of a process. But at the end of the day, it's still just a stand-alone item that performs limited tasks. They are built to service a large and extremely varied consumer base, so they are often loaded with a ton of extended features that your company probably doesn't need. Many people believe that the more features a product has, the better. This is simply not the case. In fact, these extra features can clog up your business process.

Example: Microsoft Word

Microsoft Word is a program that most people reading this book will have experience with. How many features are in Microsoft Word that you don't ever use or even know exist? Most people will only need 5% of these features, but Microsoft includes them in order to reach the largest slice of the word processing market. Since different people have different uses, they simply add every feature possible to cater to everyone's needs.

Off the Shelf vs. Custom: In a Business Context

Now let's look at these off the shelf products from a business perspective. Let's assume that a global program is designed to manage a typical business. How many features will it have that your business will never use? If you were to purchase this system, how much would it increase the training costs of your company?

You've got a new system with all kinds of features–most of which you won't ever need or use. And now you will have to train employees how to navigate around all of the stuff your business doesn't use in addition to things it will. Paradoxically, this can even make your company more dependent on people, just to make your off the shelf solution work effectively in your specific process.

On the flipside, we have the custom approach. A custom system requires you to evaluate your business processes and then build exactly what you need to operate the business efficiently. Big companies have been doing this for years because they have the resources. My point is that when you build a customized system, you're only going to include features that your business processes need.

Within that framework, you have total control over every feature that goes in. You also have control over every step of the process. Both will cut down on the amount of human interaction required. When someone interacts with this customized system, then you know exactly what will be required of them. There will be no chance to get lost in inessential features because only essential features will be included. Unlike most off the shelf products, which offer multiple ways to get to a specific endpoint, a fully customized system leads employees down only one path. This increases efficiency, accuracy, and consistency.

At this point in our culture, I would estimate that roughly 80% of the business population doesn't even know that custom solutions are a possibility. As I said before, we are well trained to look at off the shelf products first, believing that custom solutions are outside our reach. When we have a problem, the first thing we do is start looking for some sort of tool that we can buy and install to fix the problem. Who knows? It might fix that particular problem, but it fixes it in a very finite way. Furthermore, it will only fix that single portion of the business. While solving the small problem, it creates the larger problem of adding extra steps to the business processes and superfluous additional features to navigate.

The Three Reasons People Avoid Automation

Now that a better understanding of those off the shelf products has been established, it's time to dive into the excuses that entrepreneurs make in order to avoid automating their business.

Excuse #1: Money

Money is usually one of the first things that comes up. The software community has been pretty good at keeping the cost of off the shelf products low. Since they can build a single program and then sell it to millions of consumers, they are able to sell it for a very low price.

As a result, when a business is faced with a problem, these low cost solutions appear very tempting. Decision makers compare the cost of an off the shelf product (which might be $30) to a fully customized system (which might cost $30,000). Given such a big difference, they tend to focus only on the up-front costs, not even considering the important fact that the $30 product will only touch 1% of their business.

On the other hand, the custom solution might cost more up-front, but its value is almost always going to exceed that cost over the longer term through reduced labor costs, increased efficiencies, etc.

My advice to combat this excuse is to raise awareness. When analyzing possible solutions, break them down further than just the initial costs. Stop framing it as only an up-front expense. Instead, look at the overall value that it brings to your business.

To demonstrate the real value of a system, I like to build a value equation. This involves:

1. Making a list of all of the business processes.
2. Finding all of the bottlenecks (restricting points) in those processes.
3. Figuring out how automation can avoid those bottlenecks.
4. Looking at the financial reporting in those areas to get actual financial numbers related to those bottlenecks.

Once that information has been laid out, then you can work backwards from there to calculate the real financial costs of your manual processes. Once you have this report, it's easy to project the savings if those manual processes were replaced.

I've seen cases where removing these processes would save $30k a month—or even $300k per month. Having this real value staring you in the face will raise your awareness. This type of value equation is very accurate since it's based on real numbers. It doesn't require you to make speculations like "our sales are going to increase because of this automation." There's no guarantee that the automated system will increase sales. However, what is guaranteed are the financial savings from the manual systems that the automation will replace.

Once you have created this value equation, one of two things will happen.

1. The automation doesn't show a positive net return to the business within the required time frame as set by management so the cost to implement this customized system is higher than the cost over the same time frame of running the manual processes. This is rare, but does happen in either very complex automation situations and/or management setting too short (i.e. less than 6 months) of a time frame to recoup costs.
2. In most cases, the value is a positive net return. So the system might cost $300k but the savings from removing current manual processes saves more than $300k annually from operations. In this case, the expense of installing the custom system is repaid very quickly.

This is how money should be looked at when considering customized systems because if you just look at the initial cost, those cheap, one-size-fits-all, off the shelf products seem like the logical choice. By using this type of detailed reporting and analysis, you can see how much money maintaining the status quo and avoiding automation is really costing your company.

Finally, consider what those off the shelf products will really cost you. Sure, they might have a substantially lower initial cost but they're likely to cost your business a ton of money in the long-term because the software isn't doing exactly what you need it to. You're having to spend extra money on training people to navigate features that your processes don't use, the system is inefficient, and since people are involved even more in the processes, there's more room for mistakes.

It's always a good idea to convert costs into an apples-to-apples comparison. In business, looking at the ROI is always a more desirable approach than simply comparing initial costs.

Excuse #2: Lack of Knowledge

A lot of the time, people don't even know that automation is a possibility for certain processes. This also falls back on the fact that we are trained to look for these off the shelf products as a solution. This causes people to instinctively look for a better off the shelf product rather than a customized system because they simply don't know there is a better way.

Let me ask you a question: how many times have you heard people say that they bought a new product, installed it in their company, and then complain that they hate it because it's not doing what they need it to? The sales guy promised all sorts of potential, but in reality it's only doing about half of what's needed.

Of course it only does half! To be perfectly honest, if it does half of what's needed then it must be great software.

However, due to the lack of knowledge, these same people will then look for a better off the shelf product to solve the problems that were caused by the first off the shelf product. It's a vicious cycle, throwing good money after bad. Sadly, they just don't realize that there is an alternative.

The solution to this is truly understanding that there is always another possibility to explore. For example, one of my overall goals when I am looking at automating a company is to provide as much information as possible to the decision maker. You'd be surprised at how often they don't even realize that customized software is an option.

I'm not trying to bash anyone here—all of us are uninformed about something. I'm simply making a point. Technology has the potential to do so much more for businesses than most people realize. In today's world, customized systems can be created for very affordable rates that make those positive ROIs in a very short time frame very common.

Excuse #3: Lack of Time

Lack of time goes hand-in-hand with lack of knowledge. Either we don't know about these customized solutions or we don't have the time to research and discover how these solutions could work for us. In the end, it really comes down to the question, "Who's going to do it?"

Unless you're a technological guru (or your business is one that designs software), then you likely won't want to fool around with learning to build a custom system yourself. Unless you specialize in it, why would you want to? It just adds more responsibility to a CEO, who already has enough on their plate. This responsibility is also usually something they don't have experience with. It would be impossible to effectively take on this kind of task.

To say that "I can't do this on my own" is a legitimate excuse. I tend to give this excuse more weight than all of the others because it's very true. Surprised that I feel that way? I'm a business owner too and I know how challenging that can be. I truly feel that it's a legitimate excuse. As long as it doesn't stop the conversation.

The solution here is to fall back on your entrepreneurship basics: find the best people and delegate these tasks to them. Don't delegate these tasks to someone without the necessary skills to do it right. Don't delegate it to your COO or another person within your company who already has a lot of other responsibilities. Seek a specialist for these sorts of tasks.

Just because I give this excuse the most credence doesn't mean that you shouldn't try to overcome it by reaching out and bringing in the right people who understand how it all works.

This is how our business model works. We target each of these excuses and offer the company solutions. For this particular issue, we have different teams of experts who specialize in specific areas: experts on the developmental side and experts on the business process side. We don't just look at the technology involved, we look at the processes themselves.

I'm sure you've heard of the saying that "people have to work on their business, not in their business." We're that group that comes in and works on the business so that the company itself can continue to focus on working within the business.

I'm not trying to pitch my business here, I just want you to understand how we approach these automation projects so that you know exactly how to combat the lack of time excuse.

Excuse #4: Attachment

The restricting factor related to attachment is the inherent option to do nothing. That's a choice that companies make all of the time. Sometimes we will go in, do all of these value equations and lay out this whole schedule of change management so that it's perfect. It's going to take no time at all for those involved. However, even after all of that careful planning, the option to do nothing is still always on the table.

Part of this is just status quo bias again. Because your current business processes are known quantities, developed (however imperfectly) over a long period of time, they are subconsciously given preference to new, possibly risky options. Even when my team has worked the numbers and provided a logically airtight case in favor of automation, status quo bias can overrule the logic.

Another way to look at this is by using exercise as an analogy. How many people choose not to exercise, even though they know they should? If you think about it, how hard is it to put your shoes on, go outside, and run around the block for 20 minutes? We all know the value of exercise and the process itself isn't that difficult. Even so, how many people choose to do nothing? The reasoning is simply that they don't want to do it. Maybe it's slightly outside of their comfort zone or they want to sleep those extra 20 minutes every day. The comfort of routine, even when it's ultimately unhealthy, can be a powerful influence on decision making.

Turning back to a business perspective, companies often choose to do nothing, too. However, their reasoning is generally from an attachment point of view. They are attached to their current way of doing things. In this light, change can seem like a threat to stability, a potential annoyance at best, chaos at worst.

Facebook provides a great example of how this kind of attachment works. For those of you who use Facebook, look at how angry people get when the layout gets changed. People go absolutely crazy saying "I hate this new layout!" Why? Because they were attached to the previous layout. People hate change. However, let a couple of months pass and those same people who "hated" the new layout will become attached to it. Time and familiarity wins out.

It often works the same for businesses. Even when people know that it's going to be a good change and they have the money to make these outstanding changes a reality, many people will still feel uncomfortable with a new system because they are attached to their current way of doing things. They choose the option to do nothing.

Specific emotional attachments are formed with current business processes. It's been done the same way for so long that it's become part of their business culture. The fact that it's inefficient or costly might grab their attention, but it does not overpower the strength of the emotional attachment.

A second form of attachment that can influence business decisions revolves around the people being affected by the initiative. More often than not, an automated system is going to affect people in the company. It might force these associates into different roles or sometimes out of the company entirely. Depending on who's making the decision and their emotional connection to these people, they may choose not to go through with it.

Let's look at another example that I like to call 'Judy the Bookkeeper.' Judy has been with the company for 30 years. And in her own way, she is holding the company hostage. Everything is dependent on how quickly she produces reports and how well she operates, and the whole company has become dependent on Judy. This is surprisingly very common in larger companies.

People love Judy the bookkeeper. She has been there forever and she's loyal and friendly to everyone. Who cares that it takes her six weeks to produce a month end report? We love Judy!

That's how people think. We become so attached to our colleagues that we sometimes sacrifice logic and smart business practices. Even in the face of all of the facts, value equations, and long explanations of how an automated system would be more efficient, they will still choose Judy. What is it that connects us to other people this way? As was the case with status quo bias, there is an evolutionary explanation for this behavior. Over many thousands of years, humans evolved to closely bond with members of their "tribe" or, as psychologists call it, their "in-group": the people they share values and goals with. Judy has been working (no doubt to the best of her abilities) to make our company a success, just like us. So even when her performance is hindering growth, we have a hard time seeing her faults for what they are.

When you get into these situations sometimes logic will not prevail over attachment. When facing a decision, many find themselves stuck on the idea of what's going to happen to Judy the Bookkeeper. When this happens, you have to decide consciously whether or not Judy is worth a million dollars.

To be sure, there are entrepreneurs that will say yes. They will say that they make enough money as it is. They will be okay with their current system and believe it to be worth a million dollars to not rock the boat or shake up the longstanding order.

In this situation, all I can do is challenge people to make their decision consciously, with all the facts and options on the table. I ask them to look at me and say: "I know it's going to cost me a million dollars, but I don't care." No excuses, no rationalizations. Further down the road, there may be a situation that forces them to go down the path of automation. It might stop being a simple profit issue and become one of survival. In that case, when decisions are forced upon you and your own security is on the line, emotional attachments often fall away. But provoking a crisis by inaction is not the best way to reach a decision.

Whenever anyone is talking about a big change in their lives or company, it's always going to be disruptive. It's always going to feel awkward as we acclimate to a new situation and it's likely that we'll feel negative emotions along the way. So it's important to look at the big picture and where your company is going. Looking at the bigger picture will help you reach those goals. That ROI is going to be out there. That ROI in the long-term is worth much more than any of these short-term issues that you may fear. Just remember that the fear of all of those things is always way worse than the reality.

Let's imagine that you are running a marathon. You're standing at the starting line, the race course stretching out for miles in front of you and the run ahead feels daunting. Once the whistle blows, you begin your long journey toward the finish line. Sure, it might get tough at times, but every step you take will get you closer to the goal. And once you finally finish, you will realize that the effort you put in was worth every moment.

That's exactly the way you should look at automation. No matter how daunting it seems at the beginning, when you get through it and look back at the journey, you will realize that it wasn't so bad. It was worth every moment.

4 THE MINDSET OF AUTOMATION

Whoever is making the decision to automate a company has to be in the correct mindset. This mindset for automation is comprised of three key categories, each of which has its own subcategories. To put it in the simplest terms possible, this chapter is going to show you what a business owner or CEO (or any other decision maker) must be comfortable with if they are going to automate their company.

We use a fairly simple process when trying to identify good companies to work with. They must meet three basic requirements if they are going to automate. To get that information, we ask the following questions:

- Do they have the pain?
- Do they have the vision?
- Do they have the commitment?

Let's dissect each of these questions to see what makes the automation process really tick.

Do They Have the Pain?

What sort of pain inspired them to consider going down the path of automation? There are a lot of different types of pain. Let's start with the most obvious type of pain for a business.

Financial Pain

There are a lot of different situations that can cause financial pain for a business. Perhaps they have lost a big customer. Maybe the whole industry has changed or their margins have gotten tighter so they are forced to become more efficient.

Whatever the cause of the financial pain, the decision makers for that company are forced to search for solutions. For some, it will lead them down the path of automation.

This discussion might have been prompted by an operational finance pain, but that's not always the case. It might also be an effort to prep the company for sale. Maybe the owners are looking for an equity event. Since they are planning to sell the company, they want to make it as profitable as possible to attract buyers willing to pay the highest multiples. In most cases, automation is a great way to increase profitability while also making the whole business process easier to manage. Both are appealing to buyers.

Personal Pain

Now let's move on to another type of pain that is not so obvious: personal pain. There are personal reasons why a business owner might start looking down the automation path. Imagine a CEO who has been working 16 hours a day for the last 15 years. One day, they realize that they don't want to do that anymore: life's too short! They need to figure out how to make the company run more efficiently and have better access to information so they don't have to spend every waking moment in the office. Maybe they have reached a point where they need to be able to work fewer hours. They look at their company history and decide it's time to let the company stand on its own.

This situation actually happened to me. My business was already an automation business, so I was already focused on letting the machines do the work. But when I got the news that I was going to have a son, I knew that I was going to want to spend a lot of time with him, so it became important for me to get all my business affairs in order. Where possible, I wanted to use automation with the goal of freeing up time for my family.

Personal reasons are very common and sometimes become the pain points that drive the desire for automation.

Sanity Pain

Another variety of pain that we see often is one of simple sanity. Many times when a company grows, all the manual processes that are in place start to require more and more attention. They require constant memorization of what's supposed to be happening at any given time. You can imagine that the person having to personally memorize this stuff (often the same person making the decisions) is under a great deal of stress. Each time the company grows, they are forced to remember and incorporate even more stuff. Eventually business grows to the point where the people in charge are having a nervous breakdown because they are trying to stay on top of everything that has to be managed in the company.

This business process is not helping the company–it's hindering it. It's all revolving around one person (or a group of people) who are holding the company together. You don't want a situation like this so making a change would just be good business and mental health sense.

Risk and Competition Pain

This pain point is probably the least obvious. If the industry becomes saturated with new competitors, then prices will get driven down. The company takes on more risk since there is more competition. The margin for error that used to be wide is now just a tiny sliver. There's basically no room for error at all, because every mistake is an opportunity for competitors to move in.

So the more competition that a company has, the greater its risk. One of the best ways to cut this risk is to put an automated system in place. You eliminate the risk of error and instead bring a system on line that can easily adapt and grow with you.

Example: IPO Report

Here's another interesting example that I see quite often. As part of an IPO (initial public offering) companies have to produce a report documenting the processes that demonstrate that they have financial control within a company. Companies that have manual financial controls in place are required to check and update these reports either monthly or quarterly. Companies that have automated controls only have to produce these reports once a year. This example shows you just how accurate automation is. Furthermore, it should prove how it limits risk.

Pain of Growth and Scalability

Another pain that entrepreneurs often encounter is the pain of growth and scalability while remaining profitable. More than half of the companies that we talk to fall into this category. Here's what happens.

Generally, these are fast growing companies that started with just one or two people. They have grown rapidly to the point of having 300 employees. Growth is great. That's why we all start a business. The growth itself is not the problem, but rather that these businesses have focused so much on growth that they never updated their processes. They haven't used technology to their full advantage. They are using it as a tool rather than a unified automated system.

What often ends up happening is that the company grows, yet its profit margin remains the same. For example, when a company started with just two people in a garage, they might have made $100,000 a month. Now that it's employing 300 people, it's still only pulling in $100,000 a month. So the company is larger, yet it has not scaled up. Furthermore, there's more risk put on it now that it's large.

That's a pain point. Some companies grow in size but they do not scale upward. In fact, sometimes their profit margin goes down. The ultimate goal is to scale up and make more money—not scale down.

A big reason for this is because the company is not using automation. As they take on more clients, they have to hire more people to carry out those manual processes. With automated processes, the same company would still need people, but it would be far fewer than with manual processes. Those saved labor costs go straight to the bottom line.

Do They Have the Vision?

Once a company has found their pain point, it's time to ensure that they have the vision needed. There are a couple of elements involved in determining if a company has the vision necessary to proceed with automation.

The first is the vision of a single unified system. It's very easy for a company to have the pain and look into the automation route, but never achieve the one unified system vision. For whatever reason, they can't get past the whole out of the box tool solution. They can't envision one comprehensive system that connects all of their business processes and information into one place, giving management, employees, customers, and vendors access to real time data that they need throughout the business process.

This is the type of vision that a company must have. They don't have to have a technical understanding of how it would happen. They do need the understanding that it can happen. Whoever is making the decisions must already know that current technology makes it possible to have this one unified system at a manageable price point. That's the only way they will start thinking "outside of the box" and start considering automated solutions.

Real-Time Business Intelligence

Decision makers must also have business intelligence. When a spreadsheet is generated, all of this information that is being input has to be coming from somewhere. A leader with high business intelligence will understand that there is a way this data can be gathered as it happens. For example, at the moment when a sale happens, that data is integrated and accessible immediately. Understanding that is all real time business intelligence is about.

When being done manually, all of this data is gathered by a person or group of people and compiled in a spreadsheet manually. It takes time—often long periods of time—before it's available in a useful format.

When you build a unified system with a focus on business intelligence, you start to see all of that stuff as it happens. The advantages are obvious.

For smaller companies, this isn't really a major issue. However, as the company grows you start to see this reporting getting further and further behind. Your present decisions are being driven by old information. Reporting gets so far behind that it becomes difficult to create policies to respond to anything. They don't know what's happening until it's too late. When they finally make a change, they have to wait longer than necessary to see the results of that change. So they are either constantly lagging behind or furiously trying to catch up.

Limited Human Input

Understanding the value of limited human input is another part of the vision of automation. A common theme among small business owners is that they love to have employees. They start out with one or two people, then build a team from the ground up. As they add more people, they start quoting the size of their company by how many employees it has. It becomes a sort of badge of honor in the small business world. The number of people on their payroll becomes the defining unit of measurement for the company: not revenue growth, not profit, not market share. Quoting payroll volume in this way is a bad (and usually costly) habit!

Ask yourself this: is it better to own a company with 10 employees that generates $1 million a year in profits or to own a business with 100 employees that generates $900k a year? Seems obvious when we look at it from that perspective.

When a business starts to employ technology as part of the process, it can get a lot of work done with fewer employees. That's why limited human input needs to be part of the vision. The value of a company isn't based on how many people are employed. It's based on both net income and how easily (and profitably) operational processes are run.

Automated Controls

At last, we come to envisioning automated controls. I touched on this earlier when we looked at IPO reporting. The goal here is to reach the point where your automated system is actually operating and doing some of your management functions. Managers are put in place to hold people accountable and ensure that operations get done efficiently. That's a strategic decision that cannot be automated.

As for middle management, however, automated controls actually have reached the point where this layer of management is unnecessary. The system itself will request what it needs to keep everything running and follow up based on predetermined structures that are put in place.

When we start talking to a company and ask why they called us in, we're looking for them to explain their vision. They will usually explain some (or all) of the visions that I mentioned here. They do it without any input from us. If they don't, then they're probably not a good fit for automation.

The process is rather cool. Literally, CEOs will tell us these exact things without any prompting from us. All we'll ask is why they called us. Even though they might have a completely manual company, they will still have these exact ideas in their head.

Do They Have the Commitment?

Whoever is making the decision to automate a company must be fully committed to the process. They might have the pain and the vision, but it takes true commitment to make their automation vision come to life.

The first step is for the decision maker to understand the role they will have to play. In fact, the entire team will need to understand what measures they are going to use in determining whether or not automation is a good deal for them. By this point, they will have identified their pain. They will have a vision of where they want to be in the future. Now they need to determine if automation is really the direction they want to take to get there.

It's critical to clearly define those goals early on so that as they're designing their automation platform (or as we're designing it with them), everyone knows which goals are important. That decision making process can't be made on the fly. It's too important.

Finally, they all have to commit to that vision–everyone involved in the operation of the company. It doesn't do any good if one person can envision this automated system while the rest of the management team are happy to stick with routine, using a series of tools. They all have to have and share the commitment to the vision that's been created.

Overcoming the Choice to Do Nothing

This brings me back to another key point discussed earlier. The whole group also has to overcome the choice to do nothing. The choice to do nothing is the easiest choice and it's one that we all make on a daily basis. It's reinforced by our natural bias for the status quo and the stability it provides. You can even pull physics into the equation: "a company at rest tends to stay at rest," a contemporary Sir Isaac Newton might theorize. And in many cases he'd be right.

It's always easy to choose to do nothing, even if the result of changing is potentially good. Sometimes it takes a lot of work to get from start-to-finish. That's why commitment is key.

Willingness to Change

During the entire automation process—the design, construction, and implementation—there are going to be a lot of changes within the company. Identifying where these changes will take place is a big part of the initial planning. Who will be affected? When will they be affected? Laying out a basic schedule for the business transformation can be extremely important. It gives you an approximate idea of the time commitment and the scope of each person's job, and many times discovers things that may have been missed at an earlier stage of planning.

Whoever is in charge of the change must be committed to the entire process: it could be anywhere from 90 days to 2 years. Continuity in management and that management's commitment to the plan can make the process much less difficult.

Creating the Value Equation

A value equation is a term used mostly in sales discussions, but I like to adapt it to the automation process when planning it for a company. As mentioned before, a value equation compares the net value of using a product versus the cost of acquiring it. To illustrate, I'll share an example of how I approach this with prospects.

When my company looks at purchasing another company, we sit down and create a value equation that makes sense to us. We get an overview of the company's business processes and how they operate in the present. Then we will make assumptions of things that we could operationally change with an automated platform. Once we have all of that information, we can attach dollar amounts to the equation. That creates our customized value equation.

It makes it easy to see and compare numbers like labor costs before and after the automated processes are applied or the raw materials that are saved by improving efficiency. Those are two expense categories that are easily improved upon when you automate something. It's also easy to assign dollar values to those.

After that, we decide the multiplier that we want to use. That's where a rule of ratios comes into play. What's the ratio of return that we want? If we're planning to sell the company again, then it's pretty easy to come up with a ratio. If we increase the profitability by X amount, then we multiply it to find the value of that automation to the company when we go to sell it again. From a sales perspective, it's very clear what the ROI is going to be from applying automation to the business.

However, if you're not planning to sell the company, then you'll have to decide what type of ROI you want. If automation is going to save the company $50k a month, but the process is going to cost $1 million, then it would take 20 months to pay off the initial cost. Assuming that you can estimate the automated system will be used for at least three years, then you can see that you're going to see a positive ROI.

To be honest, 20 months is a bit of a stretch. Normally, the process works something like this:

- 90 days to implement
- 90 days to 9 months for payback of developmental costs
- 100% profit after that

That's a very quick return! So if the system will save $50k a month, then our goal is to build a system that costs $15k a month so that we see big savings right from the start.

A Willingness to Cut Labor

Finally, we come to an area where many decision makers have trouble letting go. This often difficult step is a huge part of the commitment process. Management might have created this whole value equation, where there is a vision and commitment to change, but when it comes down to cutting excess labor, attachments get the better of some people.

Here's an experience that my company faced some time ago. We were approached by a company, held many meetings, and laid out a whole plan for automation. In total, that plan would have cut labor costs to roughly the equivalent of having two employees replacing twenty. In other words, the wages of 18 employees would be saved with implementation. This would have been an ongoing savings and from a financial perspective, it made perfect sense. However, the owner was not willing to commit to it. His reasoning relied on two things we discussed earlier: attachments and the fact that he identified his success by the number of people he employed, not efficiency.

Once everything has been laid out, the company still has to sign on the dotted line. Nothing will happen without that. As with any business venture, at some point there will come a time when you have to sign the contract and put it in writing. That's the final commitment and one of the hardest things for people to do.

You can plan until you're blue in the face, but you have to eventually take action. It's just like the marathon analogy we used earlier. No matter how daunting it seems when you're standing at the starting line, you have to take that first step forward. If you're on the starting line in the first place, you must be there to run. No matter how daunting it seems or how uncertain you are, you must trust in what you're doing and feel confident that with each step you're getting closer to your vision.

If a company has the pain, vision, commitment—and is willing to sign off on the commitment—they will have a successful automation platform in the not too distant future.

5 FATAL MISTAKES THAT WILL KILL YOUR AUTOMATION PROJECT DEAD

In order for a large automation initiative to succeed, it must be approached just like any other large scale organizational change. All of the same processes and business transformation efforts are going to be required. There are seldom shortcuts.

There are going to be numerous non-technical issues that may come up during this process. When I say non-technical, I am referring to people. The two most common issues that we see among people involved in the transformation are either a resistance to change or general complacency towards assisting in the change process. The philosophy of good change management needs to be applied during an automation. Furthermore, it must be applied across the entire company–not just a few areas. This isn't always easy. Let me explain.

Normally, there are employees who will have to assist in the automation process. Sometimes, these employees can see that their own job will be significantly changed once these processes are in place. Sometimes their job will even be marginalized or potentially automated altogether. So it becomes a tricky situation.

Resistance can Kill an Automation Plan

As we've noted several times, people like routine and dislike change in their professional environments. Unfortunately, embracing change and adapting to an ever-changing set of processes and circumstances is absolutely necessary for growth and survival as a company. Leaders who are charged with the task of bringing change to organizations have many technical and interpersonal issues to address to successfully complete their task. Most of the time the biggest barrier to change within an organization is people, their commitment to the status quo, and managers attached to the stability of the existing power structure.

For the sake of clarity, let's limit the scope of our discussion to change within an organization and its effects on and response from employees and managers, not the related change within personal lives. Whenever there is an established system and hierarchy in place for a period of time, people become comfortable with their routines and friendly with their direct superiors. A particular system could remain in place for months, years, or even decades. The length of time it has been in place and the amount of resistance from people working within the system are unrelated, however. The level of internal resistance to change has more to do with the scope of change in the organization (including both short term and lasting effects), the makeup of the individual personalities involved, and the unique political structures that have formed within the system. Believe it or not, these make a larger difference than the length of time a system has been in place.

This resistance to change, of course, can be very disruptive to the change process and to the operations of the business in general. People often perform less soundly when they feel unsettled or unsure. That's why overcoming resistance to change must be done as quickly and efficiently as possible. Sources of resistance (i.e. individuals who "stir the pot") can be a change-maker's greatest foe. These resistant individuals often actively work against change through seemingly benevolent alignment with the change-makers and manipulation of the rest of the personnel base. A change-maker must correctly identify these specific "leader resistors" and turn them into allies of the change process or, as a last resort, eliminate them from the company altogether. This isn't a task any manager wants. But take automation out of the conversation for a moment: once your business decides on any course of action, how would you deal with someone working in the opposite direction? If there is no way to reorient them onto the company's path again, then the split has already happened. And it's probably for the best, ultimately.

Now, let's focus on a few general concepts that I've found helpful when trying to prevent this type of issue from killing an automation project.

Agile Management

This is a concept that comes up a lot of the time with current business management. Agile, which began as a software management methodology, has started finding its way into the business management world. It's a really smart concept, too. Agile was developed as a counter to waterfall development. Let me try to explain both kinds of development models for you briefly.

Waterfall development is a pretty traditional style model. Essentially, a person sits down and designs a system—kind of like an architect does for a new building—planning and defining every element of this new software from top to bottom. When it comes time to start working on the project, they have a blueprint that they follow step-by-step.

With the waterfall concept, you complete different steps and components of the software project until you're done. Agile, on the other hand, adds to the waterfall concept by acknowledging that there will be features that you didn't consider when you sat down to draw up the software's blueprint. Agile keeps a design process open to new things as they emerge, good or bad. There are almost always issues that come up when designing software that are not planned. Agile helps you remain flexible enough to take them into account as these issues come up.

The agile management process starts with a list of end goals. We work to understand why we are trying to achieve each of these particular goals, both long-term and interim benchmarks.

Equipped with all of this knowledge, we dive into the work itself. Using the agile management concept, we can react to whatever comes up during the developmental process because the blueprint is never set in stone; it's flexible. We are only working towards our end goals while using short-term goals as a kind of pathway to achieve those ends. The result will almost always be a better product than if we were to limit ourselves to only the waterfall concept.

This same concept has been applied to the management world. With this approach, you're able to get employees to take responsibility for the goals that are being set. They will understand why these goals are important so that whenever they test it, they can fill in the blanks as they go. This process involves human creativity where we all have to solve problems as opposed to telling everyone exactly what they are going to do and how they are going to do it.

Example: Agile Development

Even as agile development continues to migrate into business management, there is another related topic of importance called business transformation. I actually have a great contact in this field in Stefan DeVocht, who has really been pushing the whole methodology behind business transformation. It all came about by the whole agile philosophy. Stefan has worked with gigantic companies like FedEx and Shell Oil. He does the same types of business transformations that we will discuss in this chapter. However, he helps to apply them to companies globally.

We are basically taking the same concepts that large corporations have used to generate billions in revenue, and using them as a part of our automation process. And here's the cool thing about agile management. It doesn't matter if a company has 100 employees or 10,000 employees, they are all doing the same stuff. Stefan is a great resource for us because he can take concepts developed in his projects with billion dollar corporations and help apply them to smaller companies. When we start an automation project with a company, it makes the whole process much easier to implement. Plus these smaller companies can take advantage of technologies that have already helped major corporations make billions. It's a pretty strong endorsement.

Business Transformations and Related Processes

Business transformation is the application of agile management in an ongoing business operating environment. In a state of business transformation, it all falls back to the idea that events happening in the world are going to necessitate adjustments to a business. As such, people within that business have to be accountable for making these necessary transformations on a consistent basis.

From a business transformation perspective, the processes put in place within the business do not follow the waterfall model. They aren't rigid processes that people simply execute. Instead, under the agile model, individual employees are empowered with the responsibility to execute business transformation processes. In this example, the processes put in place are focused on the ability to transform the company, as opposed to a rigid structure where people just work.

Now, let's look at this from an automation perspective. One company is constantly changing–their employees are empowered to adjust within the company. They adjust the processes on their own all of the time. This type of company already has the tools in place to roll out an automated system because it's already utilizing the business transformation process. Sending a trailblazer project through this system would not be as difficult as it would be for a company operating under the rigid waterfall philosophy.

What I mean by a trailblazer project is a small component of what a company is ultimately trying to bring about. In an automation project, a company might have the desire to automate the entire company. That's great: we want to automate the entire company whenever possible. But if we want to successfully accomplish this, then several trailblazer elements will need to be created in order to begin the process. Trailblazers are those small points of entry that lead us from the beginning to the end. They demonstrate that you can get from point A to point Z. All of these elements are actually places to roll out the broader changes in automated processes.

What the trailblazer offers is a chance to create a business process transformation within the company environment. You create a concept of operation that serves as the plan allowing you to continually adapt to changes. Instead of building rigid processes, you will be using the concept of operations to document each of the steps that are necessary to create constant transformation within a company. And this brings us to another concept known as *"fail fast."*

Fail Fast

Fail fast is an approach to developing a product or running a company that embraces lots of experimentation with the belief that inevitably some will succeed and grow, and some will fail—but preferably in a rapid and informative way. The idea of fail fast is important, as it ties back to the overall agile methodology. You've probably heard someone say "you want to live fast." In other words, it's better to find out if something doesn't work now versus six months from now. The value with fail fast is that you learn by experience if something doesn't work and have a better understanding of how to improve or fix it. The idea is that you're better off just trying out ideas for solutions while assuming that the first attempt is most likely going to fail.

Within your normal transformation operation, you will need to assume that failures are going to be a part of your company culture. No process is perfect: you know that there are going to be issues that come up. The great thing about discovering a failure is that it gives you the opportunity to improve. Approaching each failure as an opportunity to learn, experiment, and adapt provides the method for finding ultimate solutions. So the concept of fail fast is important to the entire transformation process. The more you try new things, the more of these small failures you will encounter. Discovering and fixing each of these small failures brings you a step closer to achieving your overall goals.

This whole philosophy links back to the agile discussion, which tells us to just start building to see which parts work and which parts don't. Then keep adapting as you go while keeping your end goal in mind.

Listing and Solving Bottlenecks

The final part of a company's business transformation process centers on the idea of identifying and analyzing bottlenecks. A bottleneck is anything that restricts your company from infinite growth. What's the restricting factor for your company right now?

- Is it money?

- Is it too much dependency on people?
- Is a lack of technology holding you back?

Every company has these bottlenecks which show up in a variety of forms. Since there are normally multiple bottlenecks within a company, you will want to consider which bottleneck is the biggest. There will always be one at the top of the list. Once you have listed the biggest one, find the next. The idea is to build a list of bottlenecks based on priority. Whatever bottleneck is at the top of the list will be addressed first. Once that issue has been resolved, you will solve the next one when it arises.

Your business transformation should always be focused on resolving these inevitable bottlenecks. Once you can determine what it is, you can focus on creating business transformation efforts that resolve that bottleneck. Only then will you be consistently moving the company forward.

We do the same thing on the automation side. We create an end goal of complete automation. That's the true goal with any business. Then we make a list of bottlenecks that are standing in the way of that goal of complete automation. Let's assume that there are 30 people sitting at their computers and manually processing all incoming faxes. That would be a bottleneck in the process and it would need to be addressed.

This whole process is much easier when the company itself is already using a business transformation concept. Companies are constantly changing in every way, whether it's through large projects that encompass the entire company or smaller projects that only affect a few people. The fact that a company is in a constant state of change is both a challenge and an opportunity: using a business transformation concept can help you identify the positive and the negative aspects of change and then turn both of them to your advantage.

Automation will usually encompass the entire company if it is rolled out correctly, so having these mechanisms in place either before or implementing them during the automation process is a key factor in achieving success. Without approaching an automation project from this perspective, a business runs the risk of the system never being fully implemented. As we have already discussed in this book, it is very important that all of this stuff be implemented as a whole. Half measures are seldom successful and may even set you back.

These issues must all be addressed, especially for mid-sized companies that might not be using any of these formal business transformation processes. It's helpful for them to start looking closely at this stuff: not only from an automation perspective, but from a business operation perspective. These concepts are really valuable in every aspect of the company. Often, once your business starts taking these elements up in one area, applications in other areas become apparent.

Additional Thoughts

When you start thinking about specific elements of a business like automation, sometimes it's easy to overlook the broader implications. You will have an end goal and know that you need to automate processes to reach it. However, you can seldom guess the most efficient route right off the bat. You can build an offer and show people how beneficial it is, but some are going to still hate it. They like the way the things have always been done. It falls back to the same attachments and biases we've discussed throughout this book.

Business transformation works best when everyone affected by a potential change within the organization is participating in its transformation. It should be part of their daily routine. Those who are not trained to think like this are going to be preconditioned to perform their normal processes. So in order for them to truly accept changes, they need to incorporate this transformation as part of their daily routine. That's the goal. Naturally, doing it properly (and consistently) is a big challenge. The goal should be that this transformation is just a normal part of their routine. When you explain what is happening, your employees should simply see it as part of what they are normally doing. They will grow so used to being a part of the transformation process that they stop viewing it as change, but instead as the normal evolution of processes and routines.

There is a ton of information that can be found on agile management. If you're interested, just google around and further familiarize yourself with the idea. This business transformation stuff is interesting to use because, in the past, it has been reserved for gigantic businesses. Today, we're applying it across the board and the results are usually extremely beneficial.

6 AUTOMATION TODAY: EXPLORING THE CUTTING EDGE AND ITS IMPLICATIONS

So far, we've mainly approached the automation process from a practical, ground level view. This is my home territory: the work I do and the issues I face on a daily basis. It provides a more practical, granular approach to the many features of this revolutionary change to our ways of doing business. In this chapter, I want to take a step back and look at automation as it's being practiced at the cutting edge, to explore some expert opinions on what these changes may mean to human employment patterns and, finally, how society could respond.

Automation: Two Common Viewpoints

As the technology expert Richard Florida puts it, the general public (as well as many economists and other thinkers) has a split view on the future of technology and automation:

> Everyone has an opinion about technology. Depending on whom you ask, it will either: a) Liberate us from the drudgery of everyday life, rescue us from disease and hardship, and enable the unimagined flourishing of human civilization; or b) Take away our jobs, leave us broke, purposeless, and miserable, and cause civilization as we know it to collapse. (Florida, 2013)

The first view, which Florida dubs "techno-utopianism," rests on the belief that technological advances lead inevitably toward human progress and greater wellbeing. He quotes the American investor and economist George Gilder's positive vision of our automated future: "liberated from hierarchies that often waste their time and talents, people will be able to discover their most productive roles" (Florida, 2013). This techno-utopian view is widely prevalent, even among laypeople with little firsthand experience with the process itself. Approaching human history from this perspective, it's not difficult to track major technological advances, from the steam engine to the superconductor, and connect them directly to statistics on economic growth and increased productivity. This techno-utopian view, however, usually relies on a "big picture" analysis of these changes, often concealing the difficulties of individual or localized cases of employment disruption and sometimes even downplaying major social and political issues stemming, in part, from the adoption of new ways of doing business.

Counter to the techno-utopians, in Florida's view, are the "techno-pessimists," who believe that technology's influence and positive impact, while real, are greatly overstated. Techno-pessimists believe that "the low-hanging fruits of technological advance have largely been exhausted and the rates of innovation and economic growth have slowed" (Florida, 2013). An example of techno-pessimism can be found in the work of Northwestern University economist Robert Gordon, who writes:

> The computer and internet revolution (IR3) began around 1960 and reached its climax in the dot.com era of the late 1990s, but its main impact on productivity has withered away in the past eight years. Many of the inventions that replaced tedious and repetitive clerical labour with computers happened a long time ago, in the 1970s and 1980s. Invention since 2000 has centered on entertainment and communication devices that are smaller, smarter, and more capable, but do not fundamentally change labour productivity or the standard of living in the way that electric light, motor cars, or indoor plumbing changed it. (Gordon, 2012)

As I'm sure you've gathered by now, I lean more toward the optimistic viewpoint, while maintaining a realistic view of both technology's transformative capabilities and the potential economic, social, and personal implications of expanding automation. The exciting potential I see on the horizon when it comes to automation is fortified by its expanding capabilities at the cutting edge, some examples of which we'll explore next. But simply putting blind faith in the ultimately progressive nature of automation—faith without follow-up or debate on how best to transition to a widely automated economy—is irresponsible, regardless of your expertise.

Later in this chapter, we'll try to quantify to some degree the potential impact of new technology on employment, both locally and globally. We'll also take stock of some ideas aimed at easing the negative impact of automation.

The State of the Art

As we've discussed in previous chapters, most people still think of automation primarily in terms of manufacturing. We picture large robotic units with hydraulic limbs deftly assembling identical automobiles, or worker-less assembly lines snaking through automated stations on a factory floor. While these remain important examples of progress to date, they actually represent the last century's model of automation, focused almost exclusively on repetitive tasks defined by inflexible sets of programmed instructions. Today's automation avant-garde, however, is pushing these processes much further, often utilizing the increased capacity and speed of modern computing to mimic human-style discernment, ingenuity, logic, and even creativity.

Factory Robotics: the Next Generation

Thanks in large part to Hollywood—which has been offering stylized glimpses of our robotic future at least since Fritz Lang's 1927 epic *Metropolis*—the working robot, often with a human-like anatomy and quirky personality, has been a staple of the popular imagination. From *The Jetsons'* wise-cracking robot housekeeper, Rosie, to the iconic C-3PO, the "protocol droid" fluent in over 6 million languages from *Star Wars*, we've grown accustomed to imagining a future where anthropomorphic robots have taken on many of the mundane, monotonous, or complicated tasks humans must perform today.

Actual attempts to create these kinds of complex, human-mimicking robot workers, however, have tended towards limited or minor successes. Honda's much publicized android Asimo, for example, was generally viewed at best as a promotional prototype (and at worst as a gimmicky walking ad campaign), rather than a true working robot. Historically, physical working robots have tended to be both prohibitively expensive for most businesses and capable of performing a few well-defined tasks, with limited ability to adapt to new needs. New designs, like Rethink Robotics' Baxter, however, were conceived with both flexibility and cost effectiveness in mind:

> The two-armed robot has a computer-screen face with animated eyes, stands at about 3 feet, and is priced at $22,000. It is

designed to do such tasks as loading and unloading, sorting and tending of other machinery, jobs typically done by people… Baxter comes preprogrammed to do certain basic tasks, such as sorting objects. Buyers then need to adjust it to meet their precise needs, such as grasping certain shapes and moving objects in certain directions. That can be done without additional programming. (Hagerty, 2012)

What sets Baxter apart are its (Rethink Robotics uses the non-gendered pronoun when referring to its robots) price—which is comparable to that of a new company vehicle—and its ability to "learn" new precision tasks without expert re-programming (Hagerty, 2012). According to Baxter's manufacturers, a lay-person can easily program the robot right on the factory or warehouse floor:

A person can program Baxter by simply taking hold of its 11-joint arms and performing a desired task, such as plucking items off a conveyor and placing them in boxes. Baxter can also adjust to unforeseen events…If it drops an item, it stops and picks up another. If a human gets in its way, Baxter grinds to a halt instead of serving up a flying elbow. (Gillis, 2012)

Baxter's flexibility and realistic price point may make it attractive to business owners interested in "testing the waters" of automation and this may be a big step forward: while Baxter won't solve all your automation needs, even partial success will familiarize a much larger number of entrepreneurs and top-level managers with robotic technology. A business owner who has seen Baxter improve efficiency and/or lower expenses will likely be more receptive to global systems of automation. Is Baxter a tipping point technology? That remains to be seen. Its innovations in the areas of performance flexibility, ease of re-programming, and price control are certainly worthy of note.

Robot Farmers: Automation Takes a Second Run at Agriculture

Agriculture in the United States saw massive changes in the first several decades of the 20th century. New technologies such as the mechanized tractor and combine harvester, along with greater dissemination of efficient farming techniques via agricultural colleges and state agencies, pushed farm productivity to new heights (History of Agriculture in the United States, n.d.). It also drastically reduced the need for human labor—a primarily positive fact, in this case, since the nation was experiencing a sustained period of migration from rural parts of the country to cities for a variety of reasons.

The changes implemented during this first mechanization of farming have mostly endured, with continual marginal improvements to farming technology building upon the basic heavy machinery framework introduced nearly a century ago. But today a new generation of automation is coming to farming, with the promise of not only greater efficiency and lowered costs, but also minimization of environmental damage as a bonus:

> Soon, we might see fields with agribots that can identify individual seedlings and coax them along with tailored drops of fertilizer and measured sips of water. Other machines would distinguish weeds and dispatch them with a microdot of pesticide, a burst from a flame gun or a shot from a high-powered laser. These machines will also be able to identify and harvest all kinds of ripe vegetables…Robots could bring major changes…in how much energy, and thus carbon, goes into farming. (Crow, 2012)

Many of these new advances in farming technology are already being widely marketed and implemented. For example, many new tractors and combines are "self-steering, use GPS to cross a field, and can even 'talk' to their implements—a plow or a sprayer for example" (Crow, 2012). While many of these tools are virtually autonomous, most still leave room for human operators and follow traditional farming techniques. Most likely, this will not be true when the current wave of farm automation is complete:

> When fully autonomous systems take to the field, they will look nothing like tractors…lightweight robots could remove the need for plowing altogether…[and] precision spraying system[s]—based on the inkjet printer—can apply microdots of herbicide to the leaves of weeds…cut[ting] chemical use by up to 80 percent. (Crow, 2012)

The benefits of this kind of automation are pretty obvious: lightweight robots will do less damage to the soil (current plowing techniques are mainly used to undo soil damage caused by compaction by heavy machinery). They will also require less energy to operate and reduce the volume of both water and herbicide needed through precision watering and weeding of individual plants, not entire fields. What's most interesting to me, however, is how these new technologies are being applied to an industry that was already heavily automated over the last century for traditional reasons: improved productivity and reduced labor costs. As we can see, the newest technologies in this industry take these goals for granted, while tackling more specialized problems such as environmental pollution and water/energy/soil conservation. The reactions of farmers are also of interest. As one developer of agribots drily observed, "When we started on robotics in the mid-90s, growers were laughing and skeptical…but when we demonstrated a cucumber harvester they asked if they could buy it tomorrow" (Crow, 2012).

Agribots are already seeing wider implementation in Japan, a country with a limited and aging labor supply and an agriculture sector highly reliant on imports (Dorfman, 2009). The Japanese government has even gotten involved, funding demonstration projects and targeting a ten percent reduction in food imports over the next decade thanks, in part, to robotic farming (Crow, 2012).

Robots that Learn Like (and From) Humans

As we saw in the case of Baxter, the factory/warehouse robot that can be intuitively programmed for specific tasks by non-expert humans, the ability to "learn" new tasks is a big part of what sets cutting edge robots apart from their rigidly programmed predecessors. Innovative notions of robotic control are exploring the process of skill "learning" via motion capture techniques, which "record the motion of a human subject and map it to the kinematic structure of a robot" (Babic, Hale & Oztop, 2011). These robots are programmed with the capacity to "observe" inputs from a human controller via motion capture technology with the goal of developing "an autonomous controller derived from the human's performance" (Babic, Hale & Oztop, 2011).

Ingeniously, the robots in these studies are programmed to observe and learn human motions and tasks, but are not pre-loaded with instructions for how to perform any specific actions. By observing the posture and angle of human body parts, measuring movement parameters and quantifying the force application needed to perform a task, the robots themselves internalize this data as part of their core functionality without the need for direct programming. Essentially, this is a technological version of follow-the-leader, with human subjects performing a set task which the robotic "observer" then integrates and mimics (Babic, Hale & Oztop, 2011).

Perfecting this sort of robotic learning system could be extremely valuable because the human ability to learn and execute movement within three dimensional space is far superior to machines. The amount of coding needed to create even a rudimentary equivalent of what humans do intuitively is prohibitive. Instead of reinventing the proverbial wheel, this technology actually mirrors a human's sensorimotor learning capacity, capturing a root behavior from an example and then "building an autonomous controller so that the robot can perform the task without human guidance" (Babic, Hale & Oztop, 2011). Experiments to date have proven this technique feasible, at least with basic reaching and balancing skills. Much more research in this area will be needed to expand these robots' capability to more complex movements or processes. But the paradigm challenging nature of these experiments is clear. And as we'll see in our third example of cutting edge robot technology, similar "learning" processes can also be applied to the intellectual work of scientific research.

Birth of the Robot Scientist

Since the dawn of the digital age, human scientists have been pioneers in utilizing modern computers' ability to quickly and accurately analyze and perform complex calculations with data sets too large or complex for humans to efficiently complete. Computing capacity has been critical to the development of our modern cosmological understanding of the expanding universe; the genetic sequencing of human and animal DNA; the development of new pharmaceuticals and medical treatments; and countless other scientific breakthroughs.

Today, however, scientists are stretching robots' contribution to science from straightforward calculation problems to actual computer-generated hypotheses about experimental data, as well as the design and execution of follow-up experiments—elements of the scientific method that were exclusive to human actors until now:

> Computer technology for science has been steadily improving, including "high-throughput" laboratory automation such as DNA sequencing and drug screening. Less obvious are computers that are automating the process of data analysis and that are beginning to generate original scientific hypotheses. In chemistry, for example, machine-learning programs are helping to design drugs. The goal for a robot scientist is to combine these technologies to automate the entire scientific process: forming hypotheses, devising and carrying out experiments to test those hypotheses, interpreting the results and repeating the cycle until new knowledge is found. (King, 2011)

Researchers base their programming for these robot scientists on logic statements: well established forms for describing knowledge more precisely than common language and capable of generating logically valid extensions of that knowledge. Computer programming has relied on logical forms to translate information into formats recognizable to computing systems since the days of Alan Turing. By integrating large stores of background information on scientific problems within this logical framework, the ability to automatically perform physical experiments, and the capacity for analyzing and integrating experimental data back into theoretical structures, robot scientists are solving scientific problems whose complexity have frustrated human scientists for decades.

Adam, a room-sized robot laboratory and hypothesis generator in one, has been studying the problem of how a certain yeast strain uses enzymes to replicate and produce waste products (including alcohol) useful to humans. Equipped with robotic arms, liquid handling devices, centrifuges, incubators and freezers for sample storage, Adam is capable of "initiat[ing] about 1,000 strain-media combinations a day all on its own," says one of its developers, Ross King (King, 2011). Utilizing the tireless nature and capacity of modern computer processors, Adam can sift through vast amounts of experimental data, identify potentially useful information using its database of relevant scientific facts, and even make hypotheses about the meaning of results:

> Given a set of hypotheses with associated probabilities and a set of possible experiments with associated costs, the goal we set for Adam is to choose a series of experiments that minimizes the expected cost of eliminating all but one hypothesis. Pursuing this approach optimally is computationally very difficult, but our analyses have shown that Adam's approximate strategy selects experiments that solve problems more cheaply and quickly than other strategies, such as simply choosing the cheapest experiment. In some cases, Adam can design one experiment that can shed light on many hypotheses. Human scientists struggle to do the same; they tend to consider one hypothesis at a time. (King, 2011)

Adam eventually generated and experimentally confirmed 20 unique hypotheses during the yeast studies described above. His human developers and scientific partners then manually confirmed Adam's conclusions, finding only one conclusion that appeared wrong, seven that were correct but already known via other research (which Adam didn't have prior access to), and twelve that appear to be completely new to science (King, 2011)!

So the Future is Now?

Looking at these examples, I think two trends seem clear:

1. Automation is catching up to the popular imagination with new developments lowering the price of entry and offering greater flexibility. In industries like agriculture, which are already highly automated, the new generation of robots are approaching

maximum automation, responsible for cultivation from seed to harvest.
2. The ways robots are learning and performing their work is becoming easier, faster, and more human. These robots are able to learn more or less on the job, utilizing motion capture and video camera technology. And the speed and computing power of modern processors and networking can provide deep stores of background information to these robots as they work, allowing them to mimic the scientific method and other formerly human-only processes.

That these developments will significantly impact humans in the near term—economically, socially, even psychologically—seems more than likely. But have we, as a species in potential flux, begun to properly grapple with this probable future?

Not nearly enough, I would contend. However, a growing amount of scholarly work is being done in this area. In the next section, I'd like to briefly review some relevant ideas on automation and its impact on humans at all levels of society. While this is by no means an exhaustive review of what's out there, it may provide an entry point for thinking about these issues and provide direction towards areas you may want to explore in more detail.

Technological Unemployment: Historical Attitudes and Modern Concerns

From historical giants like John Stuart Mill, Adam Smith, David Ricardo and Karl Marx to Industrial Age theorists like John Maynard Keynes, Thomas Schumpeter and Paul Samuelson, economists, philosophers and other experts have been looking at the impact of new technologies on labor for centuries. Specifically, these experts have speculated on how an economy re-absorbs people whose jobs are replaced by machines.

The technical term for jobs that are made obsolete by new technology, straightforwardly enough, is: technological unemployment. Traditionally, it has been widely argued that with each new wave of labor displacing technology, an equal or greater number of new jobs are created. These claims were summarized first by the "Law of Markets" proposed by the French economist Jean Baptiste Say in the 1800s: "A product is no sooner created, than it, from that instant, affords a market for other products to the full extent of its own value" (Say, 1834). In other words, to borrow a common expression of this concept: "supply creates its own demand." As modern economic theory developed, a number of critics challenged this assumption, perhaps most notably Keynes. But this idea that market forces (sometimes aided by government stimulus in the Keynesian model) would create landing places for those displaced by new technology has persisted, often amplified in the work of the free market's most forceful proponents.

While this tendency toward eventual labor equilibrium does hold some historical water, the timeframe and "best practices" for regaining equilibrium have varied greatly from one technological unemployment crisis to another. More worrisome perhaps, a number of contemporary economic experts have raised questions about whether the information technology revolution might be changing the way technological unemployment is cleared by normal market forces. As the economic theorist and international policy advisor Jeremy Rifkin noted at the very beginning of this period in 1995:

> We are rapidly approaching a historic crossroad in human history. Global corporations are now capable of producing an unprecedented volume of goods and services with an ever smaller workforce. The new technologies are bringing us into an era of near workerless production at the very moment in history when population is surging to unprecedented levels. The clash between rising population pressures and falling job opportunities will shape the geopolitics of the emerging high-tech global economy well into the next century. (Rifkin, 1995)

As computers and other forms of business automation have become globally indispensable in the intervening years, even those who benefited most from these sweeping technological changes have speculated that robots (whether physical or software) may be a new kind of challenge to human labor. A 2014 *Business Insider* article noted tech icon Bill Gates' somewhat cautionary prediction:

> Software substitution, whether it's for drivers or waiters or nurses … it's progressing. …Technology over time will reduce demand for jobs, particularly at the lower end of skill set…20 years from now, labor demand for lots of skill sets will be substantially lower. I don't think people have that in their mental model. (Bort, 2014)

Speculation, even by technology experts like Bill Gates, is no substitute for quantitative analysis, however. So what do the numbers say about technological unemployment in the information age? Are we navigating another typical phase of labor displacement and eventual re-absorption? Or have computers made it a whole new ballgame?

Automation, Computers, and Unemployment: By the Numbers

In 2014, researchers from the highly respected Pew Research Center canvassed a group of 1,896 technology experts (selected for their experience, influence, and input in the economic or scientific literature) and asked a fairly simple question:

> The economic impact of robotic advances and AI—Self-driving cars, intelligent digital agents that can act for you, and robots are advancing rapidly. Will networked, automated, artificial intelligence (AI) applications and robotic devices have displaced more jobs than they have created by 2025? (Smith & Anderson, 2014)

Their results, tellingly, were almost an even split: approximately 48% of experts polled believed that by 2025 "robots and digital agents will have displaced significant numbers of both blue- and white-collar workers," with many concerned about major impacts on society. The remaining 52% of experts predicted just the opposite: that, while these technologies will no doubt displace a number of human workers, "human ingenuity will create new jobs, industries and ways to make a living" (Smith & Anderson, 2014). This fractured opinion among experts is nothing new. And as we'll see, even researchers working with comparatively solid numbers have a hard time offering predictions on their impact.

In 2013, the Oxford-based economists Carl Frey and Michael Osborne released one of the first comprehensive studies of the question of recent technological progress and its impact on specific sectors of human employment. The study looked at 702 specific job categories and, as the researchers explained in their forward, their methodology was novel:

> Needless to say, a number of factors are driving decisions to automate and we cannot capture these in full. Rather we aim, from a technological capabilities point of view, to determine which problems engineers need to solve for specific occupations to be automated. By highlighting these problems, their difficulty and to which occupations they relate, we categorise jobs according to their susceptibility to computerisation. The characteristics of these problems were matched to different occupational characteristics...allowing us to examine the future direction of technological change in terms of its impact on the occupational composition of the labour market, but also the number of jobs at risk should these technologies materialise. (Frey & Osborne, 2013)

Frey and Osborne's conclusions were sobering. Using the methods above, they predicted that "about 47 percent of total US employment is at-risk." Their model predicts job losses in many of the repetitive, "non-cognitive" manufacturing jobs that have been vulnerable to automation for the last hundred years. But they also see whole job classes based around "non-routine tasks," once believed safe from automation, becoming almost entirely obsolete in the next several decades:

> While the computer substitution for both cognitive and manual routine tasks is evident, non-routine tasks involve everything from legal writing, truck driving and medical diagnoses…In the present study, we will argue that legal writing and truck driving will soon be automated (Frey & Osborne, 2013)

Even armed with these results, though, Frey and Osborne concede that there isn't a sound method for predicting how or whether displaced workers will be re-integrated into the labor market:

> Technological progress has two competing effects on employment (Aghion and Howitt, 1994). First, as technology substitutes for labour, there is a destruction effect, requiring workers to reallocate their labour supply; and second, there is the capitalisation effect, as more companies enter industries where productivity is relatively high, leading employment in those industries to expand…Although the capitalisation effect has been predominant historically, our discovery of means of economising the use of labour can outrun the pace at which we can find new uses for labour as Keynes pointed out (1933). (Frey & Osborne, 2013)

So even after a deep look at more than 700 jobs, analysis of the current state of technology, its development trends and their potential impact, Frey and Osborne acknowledge that their crystal ball is also foggy. They do, however, put forward some recommendations for how governments might ease the coming dislocation and for how individuals should approach their personal approach to job transition. We'll touch on these in a little bit. First, let's take one more quantitative look at this issue, this time focused on a major European economy that has seen automation and robots become very pervasive in a number of industries.

The Italian Job: a Technological Unemployment Case Study

Riccardo Campa is a sociologist and researcher working at the University of Cracow. Campa's 2014 study "Workers and Automata: a Sociological Analysis of the Italian Case" looked at the impact of automation on the Italian labor market. As he explained in the study's abstract:

> Italy is Europe's second nation and the fourth in the world in terms of robot density, and among the G7 it is the nation with the highest rate of youth unemployment. Establishing the ultimate causes of unemployment is a very difficult task, and the notion itself of 'technological unemployment' is controversial. Mainstream economics tends to relate the high rate of unemployment that characterises Italian society with the low flexibility of the labour market and the high cost of manpower. Little attention is paid to the impact of artificial intelligence on the level of employment. With reference to statistical data, we will try to show that automation can be seen at least as a contributory cause of unemployment. (Campa, 2014)

Like Frey and Osborne, Campa acknowledges that the subject is tricky, but by confining his study to one country and comparing the increasing penetration of computers and robots in Italian manufacturing to the shrinking percentage of workers in that sector, he concludes:

> ...during the last thirty years, a trend has emerged that is characterised by a fall in the number of industrial workers and an increase in industrial output...The other factor that grows noticeably during this same period is precisely automation, that is, the massive use of computers and robots in industrial manufacturing. (Campa, 2014)

Proving direct causality here would be very difficult, but Campa doesn't believe that causality must be firmly established before interventions are considered:

> If it is true that the worker or the employee that the machine has replaced can find another job, perhaps a new kind of job, it is also true that they might not have the skills required for the new job (for example: computer maintenance) and that, in order to acquire them they will need months and perhaps years –that is, if they are successful...if one wants to avoid the instantaneous collateral effects of technological unemployment, one will have to

play the public hand in addition to the invisible one. (Campa, 2014)

Campa offers some suggestions to the "public hand," including reducing working hours, providing greater social welfare benefits, or even instituting a Basic Income Guarantee (BIG), which would keep workers afloat financially as they transitioned to new sectors. As he summarizes his position: "the best solution is not banning AI, but rather implementing social policies that would permit…the benefits of robotisation and automation without the unwanted collateral effects of unemployment or increasing job precariousness" (Campa, 2014).

Frey and Osborne also propose a potential public response to technological unemployment, urging skills development:

> Development strategies…ought to leverage the complementarity between computer capital and creativity by helping workers transition into new work, involving working with computers in creative and social ways. Specifically, we recommend investing in transferable computer-related skills that are not particular to specific businesses or industries. Examples of such skills are computer programming and statistical modeling. These skills are used in a wide range of industries and occupations, spanning from the financial sector, to business services and ICT. (Frey & Osborne, 2014)

While Frey and Osborne don't necessarily endorse Campa's regulatory or redistributionist suggestions, some of those ideas have been gaining traction in surprising places. Let's briefly look at the question of a Basic Income Guarantee in the context of modern technological unemployment and how it has attracted supporters from the left, right, and center. Could BIGs become the big idea for easing human disruptions caused by automation and robots?

The Question of Basic Income

The concept of the Basic Income Guarantee isn't new: American founding father Thomas Paine proposed a version of this idea all the way back in 1795 (though he was not the first) and different models for providing basic income have been promoted by thinkers as varied as the British philosopher and nominal socialist Bertrand Russell and the libertarian economist Milton Friedman (Basic Income, n.d.). The underlying concept is pretty simple: as an addition to or replacement for other forms of social security, a Basic Income Guarantee would provide an unconditional transfer of money to citizens which they could use to guarantee their basic needs. Schemes for funding BIGs generally focus on either increased taxation of private industry or, in proposals from right-leaning thinkers, by abolishing other forms of welfare and/or streamlining the tax code and utilizing the savings to fund the program.

The concept's popularity has fluctuated historically, but has often resurfaced in periods of technological or structural unemployment as an "outside the box" solution. Recently, as concerns about contemporary unemployment related to increased computerization and automation have become more prevalent, a number of members of the "tech elite" have endorsed the idea, including investor Marc Andreeson and venture capitalist Sam Wegner (Schneider, 2015).

The sociologist James Hughes sees the particular challenges presented by contemporary technological unemployment as providing a political opening for BIGs. Hughes notes increasing openness to the idea on both sides of the political spectrum, as well as recent polling that shows a sizable and growing minority of Americans in favor of BIG-type proposals (Hughes, 2014). But despite these openings, he cautions that loose political agreements might not last once actual debates about implementation begin in earnest:

> The major drawback of an attempt at a nonpartisan coalition for BIG, however, is that, in a future with a shrinking labor force and income tax base, state revenues will have to be expanded in order to provide a livable level of BIG. Putting all the current Social Security, unemployment and disability insurance, Earned Income tax credits, food stamps and other means-tested benefits into a BIG would only provide about $5000 per adult American. Steep increases in progressive taxation, consumption taxes and expansions of public ownership would be required to expand the level of BIG to a reasonable amount. These policies are likely to be fought vigorously by the conservative wing of a BIG coalition, and face an even steeper effort at winning public support. (Hughes, 2014)

Hughes' thinking here is difficult to dispute, especially given the deeply polarized nature of political debate in the contemporary United States. That said, one also has to contend with the following, posed by the entrepreneur and "tech-elite" Andreeson, about the potential benefits of maximal automation and the opportunities for human flourishing it might create:

> Imagine 6 billion or 10 billion people doing nothing but arts and

sciences, culture and exploring and learning. What a world that would be. The problem seems unlikely to be that we'll get there too fast. The problem seems likely to be that we'll get there too slow.

Utopian fantasy you say? OK, so then what's your preferred long-term state? What else should we be shooting for, if not this?... I am not talking about Marxism or communism, I'm talking about democratic capitalism to the nth degree. Nor am I postulating the end of money or competition or status seeking or will to power, rather the full extrapolation of each of those. (Andreeson, 2014)

Basic Income Guarantees are just one of many ideas in circulation for how to deal with the impact of automation on our society. In coming chapters, I will touch on some other potential areas where changes are being proposed, specifically in how we educate young people preparing to enter the contemporary job market. As an entrepreneur in the automation business, with my own experiences and inevitable biases, I am a strong believer in the transformative power of new technology. But as the saying goes: with great power comes great responsibility. Putting some of the energy, intelligence, and creativity that drives innovation on the technical front to work on problems like technological unemployment seems like an appropriate exercise of that responsibility.

7 THE FUTURE OF AUTOMATION

Having explored the cutting edge of contemporary technology and some ideas about its growing impact on society, we'll focus in this chapter on the future of automation. What does the future look like in a world where potentially (and very likely) most processes are fully automated? It's obvious that a transformation of the world on that scale would have many lasting effects. Let's start out by discussing what it really means when we become fully automated.

Earlier in the book, I used the lighthouse metaphor to illustrate the four stages of automation. Today, our whole society is headed towards that lighthouse style of economy. Businesses will continue to transform to the point where they won't actually need people to perform their processes. Our economy will end up changing from one based on scarcity, which is how our economy works now (supply and demand). When there is a limited supply, the demand goes up. When there is an abundant supply, demand goes down. That's how it works now but I see that changing in the near future.

As automation takes more of a hold on our world, the idea of products and services becoming scarce is more difficult to imagine. We are already starting to see the major elements of this transformation since shifting to a global economy. At this time, there is someone competing in every single profession that you can imagine. The whole concept of new ideas has almost been eliminated because everyone in the world can compete with anyone else.

For example, I have a personal assistant who lives in my local area. I pay a premium for her services. However, there is no reason why I couldn't hire someone who is equally as talented from another part of the world. Our modern workload no longer really requires close proximity to be successful. My point is that there are resources available to me at very limited costs and the volume of that resource is becoming more abundant. Our interconnected world and marketplace have opened up new and creative strategies for dealing with old problems, at better prices, with much wider selection.

The Shift to an Automated Economy

Outsourcing, which lowers costs by utilizing cheaper outside labor sources rather than internal ones, has led to big changes in the way labor markets function. Although often framed in controversial terms, outsourcing shouldn't be a surprising phenomenon. It's simply business people doing what they've always done: look for ways to increase efficiency to better compete in the changing marketplace. I believe automation is the next logical extension of outsourcing. We have seen the kind of shift that has happened in the form of outsourcing that has taken place as we change to a global economy. So what kind of shift will we see when we change to an automated economy? What things are going to have to change?

Education Reform for a Post-industrial Economy

The first (and probably most important) adjustment will have to take place within the educational system. The educational system we currently have is based on an industrial revolution model. It was originally implemented (after a long period of inaction and related social unrest) in response to the huge social and economic changes produced by the Industrial Revolution. But as our economic model has shifted forcefully from its 20th century manufacturing foundations to the post-industrial "Creative Economy" of the present day, our education model has only addressed this change in fragmentary ways. As education researcher and policy analyst Daniel Araya puts it in his recent book *Rethinking US Education Policy*:

> The growing consensus today is that rising demands placed on US education are not simply rooted in a temporary economic downturn, but in a permanent sea change in the structure of the global economy…Just as twentieth century education systems formed the socioeconomic foundations for an expanding industrial society, so today twenty-first century education systems are viewed as foundations for an expanding post-industrial society. (Araya, 2015)

When I think of the structure of most 20th century industrial work, I'm actually reminded of a character created in that era's heyday in the

1960s: Fred Flintstone. As you may recall from the opening credits of the cartoon, when the crow sounds on the job site, Fred knows the work day is over. He punches out of work and drives around before going home. Fred basically does the same routine each day, without variation or experimentation. Actual tasks performed will vary, but the scope of his workday is, pardon the pun, set in stone.

Our current model for educating young people was developed to produce a Fred Flintstone-like workforce and the structure of the school day clearly show this. When the bell rings in the morning students sit down, receive instructions, perform a whole bunch of repetitive tasks, and (even more archaically) memorize lots of facts. This is actually one of the interesting quirks of our educational system. A huge focus is still placed on learning strategies that developed even before the pre-industrial revolution. You can actually trace much of it back to Roman and Greek times when people held a huge advantage in society by remembering a lot of information. If it was all in a person's brain, then there was no other way to learn it other than to seek the aid of the individual who memorized it. Books were hard to get and literacy was much less prevalent. Basically, education was focused around the idea of packing as much information as possible into individuals' brains. There was no other option for storing knowledge.

Of course, things are much different now. I can pull out my smartphone at any time, Google any topic, and be rewarded with literally millions of sources of information on the subject (arranged automatically, it should be noted, by Google's algorithm for usefulness). The whole idea of our education system revolving around packing facts into people's brains and making them sit at desks working on specific tasks is becoming less and less useful. To return to Daniel Araya's analysis:

> In the context of education, [technological advances have] made older industrialized notions of formal education increasingly superfluous…educational policies must now drive toward the new and varied needs of a computational society. (Araya, 2015)

The need for a changing educational paradigm thus seems clear. Our educational system, which has gradually lost its relevance to the modern workplace, must somehow be re-refocused, foregrounding the creative and analytical skills that make humans necessary to an increasingly automated world and allowing industrial and pre-industrial era skillsets to migrate to the history department. The best methods for making this change, however, remain open for debate. We will have to reexamine how people are educated and why they are educated the way they are. If it's not a good way to educate, then what is? And what new educational techniques does the "computational society" require? What new techniques does it make possible?

Robots & Humans: Finding Their Rightful Places

Moving forward, there are going to be two competing forces in the world: one will obviously be humans, while the other will be robots, whether in the form of computers or automation.

As we've established, there is no way that humans can compete with the processes that computers do best. This includes things like storing knowledge, complex computations, and performing repetitive tasks with unfailing precision. The fundamental problem with our current educational system is that it's equipping our children to perform tasks that, in the real world, will be performed better, faster, and more cheaply by computers.

However, humans do possess a set of skills that computers cannot compete with. These skills all revolve around creativity, which is the one characteristic that gives us an advantage over all computers. The Oxford-based researchers Carl Frey and Michael Osborne shed some light on how creativity not only preserves a place in the new economy for humans, but can also rearrange our relationship to new technology:

> If a computer can drive better than you, respond to requests as well as you and track down information better than you, what tasks will be left for labor?...Human social intelligence and creativity are the domains where labor will still have a comparative advantage. Not least, because these are domains where computers complement our abilities rather than substitute for them. This is because creativity and social intelligence is embedded in human values, meaning that computers would not only have to become better, but also increasingly human, to substitute for labor performing such work. (Frey & Osborne, 2014)

In other words, when human labor relies on our unique capacity for novel, creative thinking, automation is transformed from a literally tireless competitor to human workers into a powerful ally working towards complimentary goals. This transformation doesn't happen automatically, however. As Frey and Osborne note:

> As technology races ahead, low-skill workers will need to reallocate to tasks that are non-susceptible to computerization—i.e., tasks requiring creative and social intelligence. For workers to win the race, however, they will have to acquire creative and

social skills. (Frey & Osborne, 2014)

Keep in mind here that creativity isn't always in the form of painting, singing, dancing, or theatre. When the word creative is used, most people jump to one of these professions. But creativity in the modern world is not limited to those tasks. Designers, inventors, and engineers are all examples of creative people, too. An engineer might use computers to speed up his computations, but at the end of the day engineers face problems that have never been seen before. They have to use their creativity to come up with a solution, meet a goal, or troubleshoot. That process is extremely fluid, complex, and non-linear: there is no easy way those types of creative processes could be automated since they involve a high level of uncertainty.

This is an example of something at which humans excel. We are great at adapting in a world that is uncertain. Computers, on the other hand, need certainty. They require information to come exactly as expected. Of course, as we saw in the previous chapter, technology is becoming more flexible. It's adapting too, but in the end, human creativity is indispensable for some tasks.

Building on our current level of technology, it will only get better. It will get faster and might even seem smarter because it can work a whole bunch of options extremely rapidly, selecting the optimum choice. Computers almost trick us into thinking they're being creative, but they aren't. They're only going through a preprogrammed list of options, using the brute force of modern computing power, and choosing the best one based on pre-set parameters.

The only way that a computer could stand on equal footing with humans would be if a form of technology was invented that simulates the human brain. Artificial intelligence like this has long been a goal of scientists and computer experts, but to date has remained frustratingly out of reach. Cognitive and computer scientist Douglas Hofstadter, who has studied artificial intelligence research for many decades, puts it this way:

> By the early 1950s, mechanized intelligence seemed a mere stone's throw away; and yet, for each barrier crossed, there always cropped up some new barrier to the actual creation of genuine thinking machine... Sometimes it seems as though each new step towards AI, rather than producing something which everyone agrees is real intelligence, merely reveals what real intelligence is not. (Hofstadter, 1979).

A solution to the problem of "thinking machines" might come in the form of quantum computing or some sort of different way of approaching the development of technology and its function. If that were to happen (and I stress the word "if") and computers did start to blend together, thinking like humans and being creative, then we will have basically created a new class of people. The ethical, moral, and societal implications are vast. But that's a debate that falls outside the scope of this book, so let's leave that one for future generations.

Judging from our current scope of technology, which uses zeros and ones to make decisions, computers now are about as far from creative as you can get. People have to foster their creativity. The adapting educational system will need to similarly foster creativity in people. In the end, individuals will need to take responsibility for learning the skills necessary to navigate this shift.

This will not be the first time this type of shift has happened in human civilization. It's happened many times, often very quickly. The growth and efficiency of technology has been on an exponential curve, while the rate at which new jobs are created has been on a linear line. This is precisely why we are seeing these jobless recoveries after the economic downturn of 2007-2008. Now, corporate profits are picking back up, but job creation is lagging behind. The reason is pretty clear. In 2007-2008, a lot of companies got into trouble because they were frantically trying to diagnose their problems, scrambling around, looking for solutions. They discovered that those solutions involved tightening their efficiency. So they began to bring in systems that could manage parts of their business in ways that were not being done before. Once the economy started to recover, these companies have seen their profits start to rise again. And since they had put those systems in place, there was no need to bring in more people.

It certainly wasn't like the old days where companies would lay people off in lean times only to rehire them once the economy stabilized. Or even prior technological revolutions, where new processes eliminated some jobs, but created many others to absorb displaced workers. Today, technology has enabled companies to bring in automated systems so that they did not have to hire more people, even to service the new technology. These systems allow companies to scale upward by simply adding more bandwidth or bringing in another server.

This can tug at our feelings of fairness or sympathy for displaced workers and as a society we should all be thinking of ways to reintegrate those workers into our economy. But it's important for individuals to be aware that it's partially their responsibility to find their place and participate in the creative aspects of being human. Those who take the initiative themselves are the ones who will become the most valuable.

The Difficulty of Predictions

I don't have a crystal ball. I don't know precisely what the economy will look like as we navigate this change. In 2013, the economists Carl Frey and Michael Osborne analyzed the potential impact of automation/computerization on 702 distinct occupations using sophisticated statistical models. Their results were startling:

> According to our estimates around 47 percent of total US employment is in the high risk category…i.e. jobs we expect could be automated relatively soon, perhaps over the next decade or two. Our model predicts that most workers in transportation and logistics occupations, together with the bulk of office and administrative support workers, and labor in production occupations, are at risk. These findings are consistent with recent technological developments documented in the literature. More surprisingly, we find that a substantial share of employment in service occupations, where most US job growth has occurred over the past decades, are highly susceptible to computerization. (Frey & Osborne, 2013)

Could half of the US's workforce really find themselves replaced by automation in the next twenty years? It's certainly possible, though the turbulent nature of technological advances (along with the unpredictable nature of government and industry responses) makes accurate predictions extremely tricky. What I do suspect is that there will be greater focus put on evolving industries like entertainment, recreation, and intellectual pursuits (ironically enough). I and many others view academic pursuit as a recreational activity. In my case, I have this view because in order to have the time and resources to participate in academic activities, all of my other basic needs must be met first. It's a luxury.

As it stands now, everyone is trained in the mindset that all one has to do is go to school and learn. In reality, you can only do that stuff when you have a roof over your head and food on the table. So this shift I keep referring to will likely give more opportunities for growth in the educational system, because people will have more resources to potentially continue learning.

As time passes, I don't know exactly what's going to happen with the way we work. Let's use an example here. We can start by assuming that it takes one million jobs to run our current economy. Of course, it's way more than a million but that's a nice, round number to work with for this example. Those million jobs require people to work five days a week. Now let's assume that 200,000 of those million jobs become automated to the point where it does not need a person running it. That's 20% of the workload gone—essentially one day each week that we would not have to work. Perhaps the five day work week will suddenly become a four day work week, giving us more time for other pursuits.

In the example above, it will have become standard to get three recreational days every week. As we move forward, I feel that this is the type of scenario that we're going to start seeing.

Is My Job Safe?

Those of you who have read this far might be worried. You may be wondering if your job is safe. That's actually a great, if sometimes frightening, question to have in mind. Let's take a look at an example to explore why and how this can be valuable.

Example: the Radiologist and the Housekeeper

Let's use two people for this example. The first went to school to study radiology and became a radiologist. He can read all of these fancy reports, scans, and schematics that are common in the radiology field. As an expert, he makes $300,000 a year to read these reports and give his highly educated opinion. This job is really important and for this reason is well compensated.

Our second subject in this example is an immigrant who has found work in housekeeping. He takes care of people's houses, keeps them clean, and organized. This man doesn't make a lot of money and is also in a field where there are a large volume of people with the same skill set.

So which one do you think has the highest risk of his job becoming automated? Take a moment to try to think through the variables involved in this question and make your best guess.

Did you guess that the radiologist is actually at a higher risk? In some ways it seems counterintuitive, but let's look at the facts. His job is much higher paying and relies on skills that have the potential to be automated. I used this example purposely because radiology is a field that has already started to see automation.

Now let's look at the whys. First of all, the basics: a radiologist generally sits at a desk. When people sit at a desk, it increases the possibility of their job becoming automated because sitting at a desk does not require a person to use their arms or legs for major tasks. Again, we don't have widely available mechanized robots with arms and legs yet, so physical work is one area where humans have an advantage over computers. If, like the radiologist, you are not using your physical attributes to complete your job, then you increase the risk of becoming automated.

Another factor is that a radiologist's work is really quite repetitive. They receive a bunch of different pictures, which might be in an unpredictable format, but they generally know what to expect. They are roughly looking for the same thing every time. Their decision is made based on colors, shadows, or whatever other qualities it takes to make the determination. The point is that it can all be documented. The radiologist went to school for eight years to learn it so it can be written down and learned from a book.

Anything that can be generically written into a book and can be done at a desk does not involve much creativity. This type of job is just executing a preset process—therefore, it's a high risk job.

Now let's look at the housekeeper more closely. He doesn't have to do any of that complicated stuff; he simply walks around cleaning (using his arms and legs) and travels to a completely new environment several times a day. Each house visited is unique, so there is no preset routine that can be applied to every house. A housekeeper must use the creative process to place his tasks in a certain order: clean each room and organize everything to conform to the unique needs of that household. So the housekeeper is surprisingly at a very low risk of automation. His job is too varied and reliant on creative solutions in the moment.

There is another key component that makes the radiologist's job more targeted for automation. That component can be summed up with dollar signs. Compare the salaries involved—$300,000. Automation experts are always looking at ROIs. So if I can automate a position that pays $300,000 by building a $1,000,000 system, then I am going to recoup my investment within three years.

Now, let's use that same money example to look at the housekeeper. To theoretically build a robot that has arms and legs to do his job would cost much more than a million dollars. And even if I were to successfully create a robot with all the skills of a human housekeeper, I only save $20,000 a year. So the ROI is obviously unattractive in this case.

It's much more valuable to focus on the high dollar target labor versus the lower paying position. We discussed McDonalds earlier in the book. A lot of that process is capable of becoming automated and the more people ask for money for those positions, the more risk they are putting on their job. With all economic decisions, if something costs more and it can be eliminated, then it becomes your first target.

Ask yourself the following questions:

- Do I use my arms and legs?
- Do I exercise a lot of creativity in what I do or am I just executing out of a book or a game plan of some kind?
- How much do I cost?

Those are three great questions for investigating if your job is a potential target for automation.

Plus, it's a great way of making that decision as a business owner. What jobs do you think your company will be looking at as far as automation goes? Small groups of people cost much smaller dollar amounts, so automation doesn't really seem justified. In this case, it makes more sense to keep using people until either technology becomes cheaper, or your business grows to the point where automation becomes justifiable. On the other hand, when large volumes of people (and their attendant labor expense) are involved, then automation starts to make sense.

The big question is: when all of this stuff happens, how can you avoid being replaced by automation? As I mentioned previously, it's ultimately the individual's responsibility to find ways of providing value in a world where robots can do the repetitive tasks. Robots will execute the game plans of the future, but people will still exclusively possess the creativity to form these game plans. Creativity got us here and it will get us past this shift and onto the next phase of automation.

8 NO MORE OPTIONS

This chapter is going to focus in greater detail on the fears that will sometimes get in the way of an individual automating their business. Sometimes a little bit of fear is healthy motivation that can help push a person in the right direction. At this point in time automation is still a luxury. While automating your business processes will give you a huge competitive advantage, it is still something that you can choose to either go through with or not. At some point in the near future, however, someone in your industry will make a huge leap and gain a significant competitive advantage. So here's your question for today: would you rather be that company that gains a huge advantage in your industry or do you want to be the company who falls behind and struggles to catch up?

As I mentioned in the previous chapter, a shift is going to happen. The final contours of this shift remain uncertain. But to refer back to our original evolutionary framework, when there is modification and selective pressure driving changes, there is no choice but to evolve or die out. Technology is quickly reaching the point where widespread automation will become standard business practice. The companies that can adapt to this technological change are the ones that are going to successfully continue to grow.

Here's a good example of exactly that type of advantage. We all know who Henry Ford is and that he's famous for automating the assembly line. Contrary to how the auto manufacturing industry had worked up to that point, Ford based his entire company on using machines in a way that everyone else up to that point was not doing. The rest of the industry was making cars completely by hand. Consider how much of a competitive advantage Ford's use of systems and machinery gave his company. Look at all of the other companies that played catch up.

This is very similar to the direction everything is headed today. We're going to start seeing a lot of Henry Ford type companies setting the bar. The more processes they can automate, the greater advantage they will have over the competition. The possibilities are truly wide open. To quote software entrepreneur turned venture capitalist Marc Andreeson, "It is literally the story of the economic development of the world over the last 200 years…Just as most of us today have jobs that weren't even invented 100 years ago, the same will be true 100 years from now." (Cain Miller, 2014)

Example: Automating Management Processes

Throughout this book, we have been talking about replacing human labor with digital labor. The process for getting this done involves replacing creative processes with labor processes wherever possible. What do I mean by that?

A lot of times when you have a bunch of divergent software tools being used to manage a company, there's a certain level of creativity that goes into operating those systems. That creativity isn't chosen for productivity purposes, instead it's only necessary because of the fact divergent software is being used and a gap needs to be filled. Let me use another example to make it easier to explain.

Our company began working with a company in Tampa that had experienced several recent mergers. Their first merge was with a company in Phoenix and shortly thereafter they merged with another company that has locations in Washington D.C., Philadelphia, and Charleston. I've been in touch with this company since they were in just two locations. They were considering automation even back then because their company is a very large commercial real estate company. They believed that elements of their operation could be strengthened by automation.

This company manages big lease deals. When a big company is going to lease out a whole bunch of office or production space, they can facilitate that transaction. The entire business revolves around managing paperwork while maintaining communication between various parties. That's what's happening on the production side.

On the sale's side, they are constantly performing heavy research to find out whose leases are coming up and grabbing the details of those companies. Then they use the information collected to reach out to those companies and offer their services. This company has a very specific process that has been proven to work for over 20 years. They know that if they follow steps A, B, and C then they will get results X, Y, and Z.

The problem is not in the processes themselves but the way they are managing those processes. They are clearly written down on paper and extremely easy to follow, but the actual management of those steps utilizes a bunch of divergent software tools to handle different parts of the process. Guess what? As we've noted before, when one person does not follow this process exactly, it fails. Everyone is aware of this fact, but it's a huge problem to solve when working with human labor. Our job is to find that solution, because there's no reason why one human weak link should be able to cause the whole process to fail.

So we work through the process as I've described it previously, looking at the business and developing potential fixes. We have a solution in mind and the guy in Tampa is excited. We're both on the same page. However, the guy in Phoenix seems a little resistant. He's not buying in. At this juncture, I started to tell a story about a similar project that we recently performed on another company that might work in their situation and the guy in Phoenix stops me.

He explains that what they really want is something that forces the sales guy to follow the procedure. "We don't need anything special. We just need to get them to actually follow the procedure,"—those were his actual words.

That example is exactly what I was talking about. There's no need to try to sell a magic bean or something. The goal isn't to change the process, but to remove the creativity within the process. Since they have a written process that they want followed to the letter, there should not be any need for creativity.

So I explained that we are not changing the process itself, but we are going to actually do it better. We are going to do it efficiently. Even though there are huge chunks of that process that can be fully automated to the point of replacing human labor with digital labor, there are always going to be places where a salesperson has to interface with a customer and do the things that they have to do. When they are doing those things, however, we're going to put processes in place that ensure that they follow the procedures exactly. By successfully creating this system, you're going to double, or even triple your efficiency. That's the bottom line.

So the plan would make it so that the sales person sits down every morning with a system that lays out exactly what they have to do as they go through their tasks. It will only allow them to proceed to the next step after completing the previous step. It's very simple in theory. There may be a million variables that have to be taken into account, but the number of variables is ultimately finite and definable. So they can be pre-programmed. By default, that sales guy will be guided into following the exact procedure with precision and no creative variation.

In this case, we're automating the management aspect of the system. Remember how earlier I noted how technology can now take over the function of middle management? This is a case in point, where the automated system, by virtue of the way it systematizes inputs, keeps the sales team accountable and on task—not by surveillance or constant reminders, but simply as an emergent property or side effect of the process itself. When there is only one way to complete a task, the task is always completed in that way.

People we encounter nearly always believe there are things that cannot be automated. And to an extent this is true–there are certain processes that will always require people to be involved in the system. When this is the case, however, the goal is to make sure that the employee is well managed and working for the system. Defining the process concretely and limiting the options for creative interpretation are key. Once you have that, everyone is suddenly working from the same page and fulfilling their role in a well-defined and efficient process.

The reason I like to tell this story is because it shows that even when I am talking with a company that knows what their bottleneck is, they sometimes still lack the perspective that allows them to work towards a solution. As close to and accustomed to their current system as they are, they often can't pick up the nuances of their problems and act on them to develop a solution.

From the example above, after 45 minutes of detailed explanations, I finally got the guy from Phoenix to buy in to the concept. It took several different ways of framing our solution, but eventually he realized that I wasn't just trying to sell him another software package. Once this resistance born from misconception broke, he opened up to the possibilities we were proposing.

Cultivating the Automation Mindset

This mind shift is important, which is why I told that story in the first place. Automation requires a broader vision than we're used to in standard business solutions, so it's important to cultivate curiosity and jettison preconceptions. There are always going to be CEOs and Presidents who are hesitant to proceed forward. Cultivating the broader perspective is the only way to break through that hesitation.

These same issues surfaced when outsourcing became an option in the U.S. The broader approach of a global market provided a huge competitive advantage to many companies. The idea of shipping stuff offshore to make it cheaper provided a huge advantage. Once those early adopters started to kick everyone's butts, those left had a choice to make. Do they embrace outsourcing, perish, or get pushed down into a smaller market? While this may seem like an easy choice, the staying power of the status quo can be formidable. Sometimes only existential crises can loosen its grip.

The television market serves as great proof of this. As of now, there are very few televisions manufactured in America. They all used to be built in America, but that changed. Cheaper manufacturing costs created a shift in the economy. This same type of shift will be seen as more companies adopt automation philosophies. Once the first pioneering group of companies completely automate, the rest of the market will either have to adapt or perish.

Put in the starkest terms: what is a choice at this time will become a necessity in the future. The things that you choose to automate will determine your success. Recall our commercial real estate business example. This company has been very successful utilizing a ton of stand-alone software systems and tools. They have been extremely smart about how they set up their processes using this type of software. I have to give them credit: they have done very well within the limits of that strategy. Even though everything is based on paper or in a software application, they create work flow and they complete it. They really have an outstanding plan in place. This company uses all of these independent programs in the most useful way they possibly can. The only step forward for this company is automation. But given their limitations as of now, they have established a great process that really works.

Deciding what things you are going to automate and when you're going to automate them is a huge step in the process. Are you going to just fill in the gaps with tools or are you going to start connecting processes so that these tools are no longer completely stand alone, linked through people? We have discussed the differences between tools and automation all throughout this book. I don't have to reiterate my position on their relative merits.

Another thing that people are afraid of is the big dollar sign that's attached to full business automation solutions. Some people just can't look past that number. Instead, they look at off the shelf tools and create a flow chart where they can just fill in productivity gaps with those tools. So they choose that option, go out and buy 15 different tools. Assuming that each one costs $100 a month, that's $1,500 a month in all. They may believe that is the extent of their expenses, but this doesn't take in account the fact that people are required to connect all of these different tools. Each person adds another $3,000-$4,000 expense. Now let's assume that this whole system requires 10 people to operate: your expenses jump instantly to $30,000 to $40,000 a month! Those software tools are in fact costing you a ton of money, but the expenses are great at hiding themselves.

That's a prime example of looking at the broader picture. Approached from our perspective, by automating a process that requires 10 people, you are saving $30,000 to $40,000 a month. If installing the automated system costs a total of $300,000, then the company will make back its investment within a year. After that, the system has completely paid for itself and the company is left with a more efficient system than before. Those monthly labor savings become pure profit. Keep in mind that this monetary number did not even take into account the fact that more sales are likely going to be generated because the system is more efficient.

It's easy to miss this when we focus only on the initial cost of installing an automated system. These are the types of things that you have to be constantly aware of, because your competitors will be too. In the near future, most of them will probably install an automated system. So you have a choice: either be the leader of the pack or the company playing catch-up.

9 A QUICK START AUTOMATION GUIDE

Automation does great things for a company so it's no surprise to see so many turning in their old processes in favor of automation. When a company finally sees the value of complete automation, there is a seven point process that's followed. This chapter will serve as a quick start guide to the automation process. As I work through those steps, try to picture how they might apply to your own business.

#1 Discovery

A company is running their operation a certain way. Everything seems to be flowing nicely. They're steadily growing and scaling upward until one day that growth suddenly plateaus. Management decides that something needs to change. Thus, the process of discovery begins. Now this is a fairly in-depth process.

Generally when a large enough company reaches the point where they need answers, they bring in experts to help them identify their problems. That allows this in-depth process to be done as efficiently as possible. Experts know exactly what they are looking for so they can gather information quickly and efficiently while not disrupting the normal operation of the company. The discovery phase is all about asking the right questions and looking at all of a company's processes to find potential areas that can be automated.

By the time an expert has finished their analysis, they will be able to determine if the company is a good fit for the automation process.

Sometimes a company might be using divergent tools, yet are small enough for this set-up to actually work. Although they could automate some of their processes to make them more efficient, they are not large enough for the cost-benefit ratio of automation to fit them. If this happens, the company should continue forward, but keep automation in mind as they scale up to the level where the cost becomes justified. Planning for eventual automation as the next step of your business growth plan will make the actual process that much smoother when the time comes.

The key thing to look at during the discovery phase is whether or not a company can get a positive ROI from automating their operations. Most companies can automate processes: that's not the question that you should be worried about. It's whether or not the cost is justified in your current situation.

On the other hand, if an expert determines that a company is a good fit for automation, then they can move on to the next step.

There are five questions that can be asked to help discover whether or not processes within your organization can offer high value results. If you can answer "yes" to one or more of these questions, you are looking at a process with automation potential.

- Do you have any "paper heavy" processes? Is there a paper form that gets routed to different process participants as part of the process?
- Do your process workers waste time looking for forms, documents, or data they need to complete a specific step?
- Does the process require manual duplication of data? Where maybe an email address or other well-defined information has to be manually copied from one system to another?
- Do your processes "hang" or seriously lose momentum because one of the process workers didn't receive an email that tells them to proceed with the next step?
- Are there other "routine" tasks that are very time consuming or can halt the process in its tracks if the task owner goes on holiday or simply forgets one day?

#2 Process and Financial Presentation

By this point, a company has discovered that they are in need of automation. They have brought in an expert who has told them the same thing with more detail. Now is when the real process begins.

A lot of notes are taken during the discovery phase and now those notes are transformed into a rough quote and outline of what the robotic system would look like. There is a rough feel for the overall cost and also an idea as to the potential financial benefits. What increases in efficiency will happen? How much will development cost? How much will implementation of automation save? Those are all questions that are answered with this stage of the process.

The expert that a company brings in will make this presentation to the company. They will provide it in a way that defines terms, explains interventions, and packages the process in a form the company can truly understand. It's important because this is an opportunity for questions to be asked and answered, hidden challenges to be revealed, and revisions to be made. Automation experts are not going to know every quirk of every little process within a large company. If they miss something in their investigations, then it can be pointed out during this phase.

This presentation is refined several times until mutual understanding of the process and agreement on its utility is reached. At this point nothing is yet set in stone, but the terms of a contract are outlined. It's important to get this started as early in the discussion as possible. It wastes time and resources to continue down a path only to discover that parties have an insurmountable objection. Take care of that now. Experts will provide an explanation of their commitment for a company to agree to before moving forward.

Financing must also be discussed now. A company will need to offer an explanation of how they plan to finance the whole project. This is standard procedure when a company goes through any major project, of course. Generally, a company will either lease through a third party or just pay the expense out of pocket.

#3 *Financial Approval*

This phase will be different for everyone, but the concept is fairly straight-forward. Projections were made in the previous step, so this is all about approval. Some experts will require a credit check while others will need to engage in more stringent due diligence. Again, it all depends on the expert providing the services and unique situation of the company receiving those services. For this phase, I will use my company's process as an example.

One thing that sets us apart from a lot of others is the fact that we offer financing. It's really easy for us because we base our decisions on the projections that we make for the company in terms of collateral. In other words, we have an accurate financial number that reflects how this automation will save the company money. Basing our financing terms on those calculations, we can safely assume that a company will be able to afford a specific payment.

The concept is the same in every situation though. This is the point when financial terms are negotiated and agreed to. If for some reason it turns out that these terms are unacceptable, then the discussion would be closed. However, it's safe to say that this is a very rare occurrence since the costs and gains of automating a company are clearly definable. At the end of the day, it's about us bringing value. We do our best to make that value evidence-based and clear.

#4 Signing the Contract

Okay, so all of that preliminary stuff is out of the way. Whew! We're almost there. With all of the information gathered in the previous three phases, an even more detailed plan is drawn up. This includes a rollout schedule, detailed design plan, agreements, payment information, and all of the steps that will be taken to perform the project. All of this information is put together into a contract. At this point, everyone has a chance to double check their numbers. There will be three different types of contracts that are generally required to begin an automation project. They are as follows:

- The Platform
- Developmental Contract
- Financial Terms

#5 Mutual Approval or Denial

With the contracts all prepared, all that's left to get started is approval. At this point, there are only two choices: either both parties move forward by signing the contracts or they pull out of the contracts. If the previous steps have been performed adequately, the case explained thoroughly, and questions completely addressed, this choice should be relatively easy to make without regret or cold feet.

#6 Automation Build + #7 Automation Installation

The final two phases of an automation project go hand-in-hand and often overlap each other. They also take the most time. While one thing is being built, another is being implemented. All of this happens while revisions come in and minor tweaks are made to the system. These two go back-and-forth a lot.

The minimum time it takes to successfully complete build and installation of an automated platform is 90 days. However, the scheduled rollout will be crafted to each company's unique situation: that will define the exact length.

For a full company-wide automation project, a rollout schedule is defined in the earlier phases. This schedule will determine how long the whole process will take. A rollout schedule is essential to ensure that a company can continue with their daily operations without disruptions. Sometimes the scheduling of implementation is the linchpin in the whole operation. It ensures that everything comes out smoothly. You might end up rolling out pieces with more time in between them than is actually needed because of some other factor that's outside of the scope of the project, but important to maintaining business functions.

For example, let's say the company wants to start a project in January but they know that Quarter 3 is always extremely busy and that they can't be doing anything during those three months. This means that once Quarter 3 comes around, the automation process will be designed to pause in order to allow normal operations to deal with the busy season. Once that period has been successfully navigated, the process will be picked up again at the start of Quarter 4.

The Importance of Ongoing Audits

Once implementation has been completed, there is a final step to the process. At several intervals after automated processes have been installed, an expert will need to audit the company. It's essential that every six months to a year, an expert walks through and does an audit of the entire system to make sure that as the company continues to operate and grow, it doesn't become un-automated by changing processes that aren't integrated onto the platform.

This sort of thing can happen very easily—sometimes almost undetectably—so it's important to watch out for it. It's very easy for a new process to begin and while employees are trying to figure it out, they fall back on old habits. Spreadsheets are created. Soon, the spreadsheets become an integral part of how the operation runs. It's the expert's job to identify when this happens and then create additional modules to pull these new processes into the automated system. The high level of efficiency achieved through automation is vulnerable to these proliferating outside processes. Ongoing audits are thus very important for the long-term viability of the project.

The key thing to remember is that when it comes to all of these types of implementations, you're not going to just do them in your spare time. Furthermore, you're not going to be able to just allow your IT guy to take on this enormous task. Having an individual within the company whose full-time job is to help oversee this change is critical. Alternately, you can bring in an outside company to take care of the whole process. I've seen companies that have had mild levels of success of trying to perform this level of automation on their own. At best, they are able to turn out some working systems. It just takes them years to get anything worthwhile.

The worst case scenario is that they end up not only failing with what they want to do on the automation side, but they disrupt their regular business operations.

As an outside expert myself, I'm no doubt biased towards bringing in a dedicated experts with the experience and resources to plan, execute, revise, and rework with efficiency. After all, this process is proven by track record. We've done it a lot of times for our companies and others.

AFTERWORD

If you've read and remained engaged with this subject matter to this point, I salute you. We've covered a lot of information and if you've gotten this far, it's an obvious indicator that automation is a subject of interest: either as a potential strategy for improving your business or simply an area you recognize will become more and more prevalent as we progress.

The subject of automation and the various related issues it brings up is truly massive. In order to focus my arguments, there are several subjects that I intentionally avoided or only touched on briefly in this book. In places, you may have questioned whether my consideration was completely thorough. I can assure you that there were many subjects that I had to restrain myself from exploring, simply to remain tightly focused. Questions about labor dislocation caused by automation and the moral implications of the changes we're seeing—subjects we touched on in Chapter 6—could be the sole focus of a book like this. I chose to guide my work here by focusing on the questions and concerns most relevant to decision makers in a business context, while offering ideas for further reading or study wherever possible.

All that said, I'd like to briefly touch on a few of these subjects and propose some ideas that may help you to further explore your own thoughts and beliefs. You may be surprised by what you actually think.

Dystopian Depictions of Technology

Our popular culture, as I've touched on in places, has shaped how we think about the growth of automation in many ways. Often in this context, the most sinister or dystopian possibilities are amplified to create suspense, horror, or other emotional responses. Think of the computer HAL 9000 in

Stanley Kubrick's film *2001* or the "rise of the killer robots" back-story of the *Terminator* film series. As you can imagine, these examples are hard to avoid when you're in my business.

Naturally, watching out for unintended dangers is just smart practice. It's built into the way I approach automation and I'm constantly on the lookout for new developments in this area. But allowing fictional representations to substitute for factual analysis is not going to clarify matters. Just the opposite. So for these reasons, I've stuck close to my own personal experience and expertise when it comes to business automation: the nuts and bolts of the process. Absent major advances in artificial intelligence, quantum computing, or other technologies, the scenarios that thrill us in the movies will remain just that: unlikely possibilities only loosely tied to real world developments. The potential dangers are also on the radar screens of many important thinkers and business leaders, including Microsoft's Bill Gates, Tesla's Elon Musk, and Google's Larry Page, to name just a few.

Philosophers, computer scientists, and other experts have also been deeply engaged in these questions at least since the time of Alan Turing's paper titled "Computing Machinery and Intelligence" in 1950. Contemporary thinkers like John Searle, Hubert Dreyfus, Susan Blackmore, and David Chalmers (to name just a few) are also integrating the technological advances we're witnessing into their own theories about technological development, artificial intelligence, and its implications. If you have a nagging desire to learn more about the subject, I'd recommend investigating further. It may not *directly* profit your business, but it will certainly broaden your mind.

Non-Biological Evolution Theory & its Discontents

Framing the growth of automation as an "evolution" is also potentially objectionable for some. Darwin's concept of evolution through mutation and natural selection has been called "history's best idea" (Chivers, 2013) because it so elegantly explains how biology has changed and progressed over time. I'm not the first person to try to adapt that "best idea" to non-biological processes. Look through the class descriptions of any college or university and you will invariably encounter evolutionary takes on economics, psychology (both human and animal), language, culture, and many other subjects. I don't claim that my approach proves that business develops according to an iron law of evolutionary principle, simply that the framework that evolutionary theory provides can help us conceptualize and better predict how businesses change over time, integrating mutations and selective pressures.

It's interesting here to touch again on Richard Dawkins and his theory of memes. While Dawkins presents memes as analogues to genes in the process of cultural evolution, he also notes one major difference: unlike genes, which mutate randomly and are selected either for or against by blind competitive pressures, the evolution of memes (and thus the culture overall) are subject to human reason and logic as a major form of selective pressure. What this means is—at least to some degree—the process is open to our intellectual tinkering and creativity. To quote Dawkins: "We are built as gene machines and cultured as meme machines, but we have the power to turn against our creators. We, alone on earth, can rebel against the tyranny of the selfish replicators" (Dawkins, 1976). What he describes here—and what I've tried to illustrate throughout this book—is really an earthshaking development: humans have the power to harness the power of evolution, the force that shaped all life on Earth, to our own ends, at least to some degree. There is great power here, but, as always, great responsibility, too.

The Moral Implications of Automation

For the most part, I have intentionally left a detailed exploration of the moral implications of greater automation out of this discussion, as well. There are several reasons for this. The first is simple expertise: I work every day in the field of automation implementation, not moral philosophy. I certainly have my own opinions on this subject, but also recognize that they are subjective and that others can reasonably arrive at dissimilar conclusions.

I have discussed how we may want to adjust our educational systems to better accommodate the realities of the modern marketplace, including automation. Because I've seen how the limitations of our current approach impact my own work, I feel qualified to offer suggestions in that area. I've also tried to survey expert opinion on the potential scope of this issue and explored some pragmatic suggestions for action. But the greater questions of how to deal with labor dislocation as a society is a subject we all need to think about and discuss. I certainly can't dictate the terms of that debate, so instead I try to encourage the culture to become more engaged. That's really the only way to reach consensus and allow solutions to emerge.

Optimistic Realism

I suppose I'm both an optimist and realist when it comes to the concerns and objections outlined above. I've seen how positively automation can impact a business, driving efficiency and productivity to staggering levels. I've also encountered stiff resistance to these advances at times. Looking closely at history, I've concluded that to some degree change is inevitable. As a true believer in human creativity (and an advocate of its liberation from antiquated ways of doing business), I believe we'll discover ways to shape that change to suit both our individual and cultural needs. But the conversation must begin now. And it must make room for the widest possible scope of opinions and ideas. It's how we've advanced this far: from the muddy, brutish struggle for existence in prehistoric human history, to our current dominance of the Earth and, increasingly, beyond.

Thank you for taking the time to investigate my personal take on this exciting journey. May it serve you well as you join the journey and begin to help shape it yourself!

REFERENCES

Adler, N. J. (2002). *International Dimensions of Organizational Behavior.* Cincinnati, OH: South-Western, a Division of Thomson Learning.

Andreeson, M. (2014, June 13). This is probably a good time to say that I don't believe robots will eat all the jobs. [Web log post]. From: http://blog.pmarca.com/2014/06/13/this-is-probably-a-good-time-to-say-that-i-dont-believe-robots-will-eat-all-the-jobs/

Araya, D. & Peters, M. (Eds). (2010). *Education in the creative economy: Knowledge and learning in the age of innovation.* New York: Peter Lang.

Araya, D. (2015). *Rethinking U.S. Education policy: Paradigms of the knowledge economy.* New York: Palgrave Macmillan.

Babič, J., Hale, J. G., & Oztop, E. (2011). Human sensorimotor learning for humanoid robot skill synthesis. *Adaptive Behavior*, 19(4), 250-263.

Basic Income. (n.d.). In *Wikipedia: the Free Encyclopedia.* Retrieved on December 23, 2015 from https://en.wikipedia.org/wiki/Basic_income

Bellamy, D., & Pravica, L. (2011). Assessing the Impact of Driverless Haul Trucks in Australian Surface Mining. *Resources Policy*, 36(2), 149-158.

Borenstein, J. (2011). Robots and the changing workforce. *AI & Society*, 26(1), 87-93.

Cain Miller, C. (2014, December 16). As robots grow smarter, American workers struggle to keep up. *New York Times.* pp. A1-A3.

Campa, R. (2014). Workers and automata: a sociological study of the Italian case. *Journal of Evolution and Technology.* From: http://jetpress.org/v24/campa1.pdf

Chivers, T. (2013). Charles Darwin: the man who had history's best idea. *The Telegraph.* From: http://blogs.telegraph.co.uk/news/tomchiversscience/100202558/charles-darwin-the-man-who-had-historys-best-idea/

Crow, J. M. (2012). Down on the robofarm. *New Scientist*, 216(2888), 42-45. From: https://www.ecs.hsosnabrueck.de/uploads/media/2012_10_27_NewScientist_Down_on_the_robofram.pdf

Dawkins, R. (1976). *The Selfish Gene*. Oxford: Oxford University Press.

Dorfman, J. (2009). "Fields of automation." *The Economist*. From: http://www.economist.com/node/15048711

Evolution 101 (n.d.). *Understanding Evolution*. Berkeley University. From: http://evolution.berkeley.edu/evolibrary/article/evo_02

FLORIDA, R. (2013). Robots Aren't the Problem: It's Us. *Chronicle Of Higher Education*, B10-B12.

Frey, C. B. & Osborne, M. (2013). *The future of employment: how susceptible are jobs to computerization?* From: http://www.oxfordmartin.ox.ac.uk/downloads/academic/The_Future_of_Employment.pdf

Frey, C. B. & Osborne, M. (2014, May 15). "Technological change and new work." *Policy Network*. From: http://www.policy-network.net/pno_detail.aspx?ID=4640&title=Technological-change-and-new-work

Gillis, C. (2012). The robot working class invasion. *Maclean's*, 125(41), 50-52. From: http://www.macleans.ca/economy/business/the-robot-invasion/

Gordon, R. J. (2012). Is US economic growth over? Faltering innovation confronts six headwinds. *Centre for Economic Policy Research*. From: http://www.cepr.org/sites/default/files/policy_insights/PolicyInsight63.pdf

Grogan, A. (2014). STANDING IN THE WAY OF CONTROL. *Engineering & Technology* (17509637), 8(12), 52-55.

Hagerty, J. R. (2012, September 18). Baxter robot heads to work: designed for factories, $22,000 machine can be taught to perform tasks." *Wall Street Journal*. From:

http://www.wsj.com/articles/SB10000872396390443720204578004441732584574?cb=logged0.7349296750035137

Hanley, C. (2014). Putting the Bias in Skill-Biased Technological Change? A Relational Perspective on White-Collar Automation at General Electric. *American Behavioral Scientist*, 58(3), 400-415.

Hecht, J. (2001). Classical Labour-Displacing Technological Change: The Case of the US Insurance Industry. *Cambridge Journal Of Economics*, 25(4), 517-537.

History of Agriculture in the United States (n.d.). In *Wikipedia: the Free Encyclopedia*. Retrieved on December 23, 2015 from https://en.wikipedia.org/wiki/History_of_agriculture_in_the_United_States

Hofstadter, D. R. (1979). *Godel, Escher, Bach: An Eternal Golden Braid.* New York: Basic Books.

Hughes, J. (2014). "A strategic opening for a basic income guarantee in the Global Crisis Being Created by AI, robots, desktop manufacturing and biomedicine." *Journal of Evolution and Technology*. From: http://jetpress.org/v24/hughes2.htm

King, R. D. (2011). Rise of the Robo Scientists. *Scientific American*, 304(1), 72-77. From: http://www.cs.virginia.edu/~robins/Rise_of_the_Robo_Scientists.pdf

Lane, N., Allen, J. F., Martin, W. (2010). "How did LUCA make a living? Chemiosmosis in the origin of life." *Bio Essays*, 32, 271-280.

Lin, P., Abney, K., & Bekey, G. (2011). Robot ethics: Mapping the issues for a mechanized world. *Artificial Intelligence*, 175(5/6), 942-949.

Metabolism (n.d.). In *Wikipedia: The Free Encyclopedia*. From: http://en.wikipedia.org/wiki/Metabolism

Palmer, D. (2009). Young Earth. In *Prehistoric Life*. (pp. 10-45). New York, NY: DK Publishing.

Paatz, S. (2010). Robot dreams. *TCE: The Chemical Engineer*, (826), 39-40.

Rifkin, J. (1995) *The End of Work: The Decline of the Global Labor Force and the*

Dawn of the Post-Market Era. New York: Putnam Publishing Group.

Samson, R. W. (2013). Highly HUMAN Jobs. *Futurist*, 47(3), 29-35.

Say, Jean B. (1834). *A Treatise on Political Economy* P138. From: https://books.google.com/books?id=WkaPTSyM8E4C&pg=PA138#v=onepage&q&f=false

Schneider, N. (2015) "Why the tech elite is getting behind basic income." *Vice*. From: http://www.vice.com/read/something-for-everyone-0000546-v22n1

Smith, A., Anderson, J. (2014) *AI, robotics and the future of jobs*. Pew Research Center. From: http://www.pewinternet.org/files/2014/08/Future-of-AI-Robotics-and-Jobs.pdf

Tattersall, I. (2010). *Paleontology: A Brief History of Life*. West Conshohocken, PA: Templeton Press.

Walker, M. (2014). BIG and Technological Unemployment: Chicken Little Versus the Economists. Journal Of Evolution & Technology, 24(1), 5-25.

ABOUT THE AUTHOR

Dan Abbate is an entrepreneur, thought leader, and investor with a career-long focus in business process automation through the use of advanced technology and organizational development and improvement. Mr. Abbate excels at leading businesses to overcome challenges and make high-stakes decisions using his experience-backed judgment, strong work ethic, and irreproachable integrity. Characterized as a visionary strategist, Mr. Abbate has a consistent record of delivering extraordinary results in growth, revenue, operational performance, profitability, systematization, automation, and continuous organizational improvement. He has a heavy transaction background including startup financing, mergers and acquisitions, and turnarounds. Between the years of 2003 and 2013, Mr. Abbate started/developed or acquired/improved and then sold eight different companies–serving various markets and industries–funded through private capital markets. He has experience in both B2B and B2C companies. A silent and lead investor in a number of technology startup companies (including www.bdex.com and www.glince.me), Mr. Abbate is also a member of the Young Professionals of Wellington, Rotary International, and Entrepreneurs' Organization of South Florida, where he is a mentor in the Accelerator Program, which focuses on guiding young entrepreneurs with companies less than $1,000,000 in revenue to achieve that milestone and beyond within two years. He enjoys participating in high-level operational initiatives, including infrastructure design, process re-engineering, turnaround management, and reorganization. Mr. Abbate holds a degree from DePaul University in Philosophy and Business, and recently presented his prophetic talk "How Not To Be Replaced By A Robot" at TedXBoca (2015). He spends his free time thinking of how to make the world run more efficiently; reading various business and industry related books and publications; running 5ks; and riding on his golf cart with his young son, Winston and lovely wife Kelly.

www.ingramcontent.com/pod-product-compliance
Lightning Source LLC
Chambersburg PA
CBHW070324190526
45169CB00005B/1739